NATURAL HISTORY AUCTIONS
1700-1972

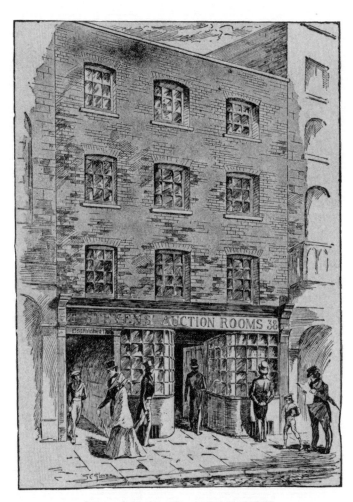

Stevens' Auction Rooms about 1840,
from the cover of a sale catalogue

TITLE PAGE VIGNETTE

'Capturing Insects', Frontispiece to the anonymous *History of Insects,* London 1839

NATURAL HISTORY AUCTIONS
1700-1972

A Register of Sales in the British Isles

Compiled by J. M. Chalmers-Hunt

CAPTURING INSECTS.

With articles by
**S. Peter Dance, Peter G. Embrey, W. D. Ian Rolfe,
Clive Simson, William T. Stearn, J. M. Chalmers-Hunt
Alwyne Wheeler**

SOTHEBY PARKE BERNET
1976

© J. M. Chalmers-Hunt 1976

PRODUCED AND PUBLISHED BY
SOTHEBY PARKE BERNET PUBLICATIONS LIMITED
36 DOVER STREET, LONDON W1X 3RB

Edition for the United States of America available from
Biblio Distribution Center for
Sotheby Parke Bernet Publications Ltd.
81 Adams Drive, Totowa, New Jersey 07512

ISBN 0 85667 021 9

All rights reserved. No part of this publication
may be reproduced, stored in a retrieval system
or transmitted in any form or by any means
without the prior permission of the publisher.

Printed in Great Britain by
A. Wheaton & Co., Exeter
and bound by G.& J. Kitcat Ltd, London

*This work is dedicated to
The British Museum (Natural History)
its Staff and Collections*

CONTENTS

Introduction	*page*	ix
Acknowledgements		xi

PART ONE Introductory Articles

Entomological Sales J. M. CHALMERS-HUNT	3
Zoological Sales other than those of Birds Insects and Shells ALWYNE WHEELER	15
Botanical Sales WILLIAM T. STEARN	20
Ornithology and Oology CLIVE SIMSON	23
Fossil Sales W. D. IAN ROLFE	32
Minerals PETER G. EMBREY	39
Shell Sales S. PETER DANCE	45

PART TWO The Register of Natural History Sales

Signs and Abbreviations	54
Catalogue Locations	54
Explanatory Notes	56
Sales 1700-1800	57
Sales 1801-1900	67
Sales 1901-1972	137
Index	181

INTRODUCTION

People have for centuries collected natural history specimens. At first, such assemblages were mostly miscellaneous in character and often originated from material obtained from ship's captains, travellers returning from abroad with curiosities for sale, or from scientific exploration. Later, with increased knowledge, natural history collections tended to become more specialised. Many were formed solely for their aesthetic appeal, some as scientific aids to research, and others as a mixture of both.

Natural history auctions first took place on the Continent in the late 17th century, and probably first became popular in Great Britain a century or so later where, in all, nearly 1600 are known to have been conducted. Many such auctions were of collections of considerable scientific significance, containing as they did type specimens, unique, rare and extinct forms described in the literature, or examples that formed the basis of important monographs. On dispersal, the sale catalogues often held the only clue to their ultimate fate.

This book sets out to present a systematic record, as far as it is possible, of every auction in which natural history specimens have been offered for sale in the British Isles and, although it may not be complete, cites most, if not all, of the more important sales. Its main purpose is to present, in easily accessible form, concise details of such collections as have been sold by auction, and to give locations of the sale catalogues whenever possible. References to auctions of other natural history objects such as books, mss. prints or paintings are included only if these formed part of a specimen sale, or if they had some special significance.

Being the first work of its kind to be published, much of the information has had to be gleaned from original sources. The main, and indeed most satisfactory of these has been the sale catalogues themselves, with the principal collections of them providing the bulk of the material contained in the *Register*. A sale catalogue's ephemeral nature however, its frailty, and the problem of classification it presents to the librarian, makes it among the hardest of documents to locate. Accordingly, searching for the specific catalogues has proved a difficult and often unrewarding task, though sometimes fortuitously bringing to light others whose existence until then was unknown. Such serendipity has furthermore brought about the discovery of bibliographical references to natural history sales of which I was previously unaware. Thus, chance has played no small part in helping to achieve the final result, but one may assume in all probability that there exist, hidden away and widely scattered, references to, as well as catalogues of, many more natural history auctions than are recorded here.

Among bibliographical aids to the compilation of the *Register*, first and foremost is Sherborn's *Where is the — Collection?* (1940, published at the author's own expense), which has contributed numerous sources of information, as has also a copy of this work, heavily annotated in an unidentified hand in the British

INTRODUCTION

Museum (Natural History). F. Lugt, *Repertoire de Ventes Publiques, 1600-1900* (La Haye, 1938-1964), though primarily concerned with art auctions, has references to a number of British natural history auction catalogues, some of which I have as yet failed to trace, including those cited in brackets in the *Register* under BSMB, C, CEB, G, P and RIA, as well as most of those stated to have been formerly in the possession of Seymour de Ricci, and now supposedly in the Bibliothèque Nationale in Paris. J. J. Fletcher (1920 *Proc. Linn. Soc. N.S.W.*, **45**: 574-576) gave the Linnean Society of New South Wales as the location of Alexander Macleay's (1767-1848) small collection of auction sale catalogues, but recently, the Secretary of the Society informed me (*in litt.* x.1974) that this is not there, and its fate is not known to me.

Two fascinating publications are Allingham's *A Romance of the Rostrum* (1924), and Horn, Kahle and Sachtleben's *Über Entomologische Sammlungen und Entomo-Museologie* (1935-1961). These record many entomological and natural history auctions, though unfortunately without giving precise dates and in the case of the former work often without any dates at all, but nevertheless furnishing particulars of a number of sales previously unknown to me.

Then there are J. R. le B. Tomlin, Shell Sales I-III and IV (in *Proc. Malacological Soc.*, 1941-1942, 1949); and the much-appreciated unpublished manuscript of F. J. King, *English Non-book Auction Sales, 1801-1884* (details of which were extracted by M. J. Rix), which refers exclusively to catalogues in the Bodleian. Finally, I must not fail to mention *The Athenaeum* (London, 1828-1921), a periodical that advertised numerous natural history auctions, particularly those conducted by the firm of J. F. Stevens.

Original Manuscript of the *Register*

I have deposited this in the Manuscripts Department of the General Library, British Museum (Natural History). It is in eleven notebooks, of which the last consists of an appendix of about 500 numerical notes arranged chronologically under five alphabetical serials from A to E. Many of these appear herein as footnotes, and a number is incorporated in several of the contributory articles, but there are others of some interest that are not published in the final work.

Request for Information

Owing to the difficulties in locating many old sale catalogues and almost all references to relevant provincial auctions, the *Register* cannot be regarded as in any way exhaustive. With a view to issuing a supplement, I would be glad to be notified of any auctions not referred to in the *Register* with, if possible, locations of the catalogues. I would also be grateful for information about any existing unlisted catalogues of sales recorded in the *Register*. Although every effort has been made to ensure accuracy, mistakes may occur and I would welcome information about any errors that may be found.

ACKNOWLEDGEMENTS

One of the most pleasing aspects of this undertaking has been the kindness shown to me in my search for auction catalogues, information about auction sales and particulars relating to the collections sold. My thanks are owed to a great many helpers and correspondents, and I now express my gratitude to one and all, though lack of space prevents me from naming individually many of those to whom I am indebted.

It is difficult to make distinctions when help and encouragement have been so freely given, but I do especially thank the undermentioned:

Mr D. B. Janson and his father, Mr R. B. Janson, for permission to examine and record from their fine collection of auction sale catalogues; also, the former for his keen interest and sound advice on many points relating to the *Register;* the staff of the British Museum (Natural History), South Kensington - in particular, Mr R. E. R. Banks for his constant interest, ever willing help and efficiency at dealing with my many requests; Miss Pamela Gilbert for kindly granting me the use of her unpublished *Compendium of Biographical Literature on Deceased Entomologists;* and Mr R. I. W. Atkins, Mr. B. J. Clifton, Mrs Ann Datta, Mr P. Embrey, Mr M. Halliday, Mr A. P. Harvey, Mr M. J. Rowlands, Mr K. G. V. Smith, Dr W. T. Stearn, Mr Alwyne Wheeler and Mr P. J. P. Whitehead, for their valued assistance and helpful suggestions.

I am also grateful to Mr L. Christie, Mr S. Peter Dance, Mr Eric Gowing-Scopes and Mr R. Nichols, for the loan of, and for permission to record from, their collections of auction sale catalogues; to Mrs Audrey Smith (Hope Department, University Museum, Oxford) and Dr Hugh Torrens (Department of Geology, Keele University, Keele, Staffs.), for much help in tracing catalogues; also, Mr M. F. Stephens and Mr I. F. Lyle (Royal College of Surgeons), Mr M. J. Rix (Bodleian, Oxford) and Mr J. H. van Haeften (Messrs. Christie & Manson, Auctioneers) for help of a similar nature.

My thanks are also due to Miss Jane Boulenger and Mr Philip Wilson, who kindly accorded me facilities for studying and recording from Messrs. Sotheby's marked set of their own sale catalogues.

The staffs of the Reading Room, North Library, Department of Ethnography, Department of Coins and Medals, and Department of Prints and Drawings of the British Museum, London, for help in locating catalogues; likewise, the staff of the Library of the Victoria and Albert Museum, London.

I also desire to acknowledge gratefully my indebtedness to the following for information or for assistance in tracing sale catalogues: Miss J. A. Agnew (City Museum and Art Gallery, Birmingham); Mr W. H. Barrow (Queniborough, Leicester); Mr C. W. Bayliss (Asst. Librarian, Geological Society, London); the late Rev H. E. Biggs (Bromley, Kent); Dr A. Blok (Rottingdean, Sussex); Mr P.

ACKNOWLEDGEMENTS

J. Bloomfield (Keeper of Natural History, Bolton Museum); Mr. G. Bridson (Librarian, Linnean Society, London); Dr C. Helen Brock (Department of the History of Science, University of Glasgow); Mr Martin Dawson Brown (Acaster Selby, York); Mr T. E. Crowley (President, Conchological Society); Mr P. Doughty (Keeper of Geology, Ulster Museum, Belfast); Miss Pamela Dutton (City Library, Liverpool); Messrs. J. & R. Edmiston (Auctioneers, Glasgow); Mr I. M. Evans (Keeper of Biology, Leicester Museum); Dr V. A. Eyles, (Great Rissington, Glos); Miss Pamela Eyres (Librarian, National Gallery, London); Dr J. A. Gibson (Kilbarchan, Renfrewshire); Miss Ginsberg (Librarian, National Portrait Gallery, London); Miss C. W. E. van Haaften (Rijksbureau voor Kunsthistorische Documentatie, The Hague, Holland); Mrs B. P. Hall (Asst. Editor, *The Ibis*); Mr D. Heppell (Royal Scottish Museum, Edinburgh); Mr J. E. T. Horne (Secretary, Mineralogical Society); Mr R. Hughes (Librarian, Department of Zoology, University of Cambridge); Mr Charles MacKechnie Jarvis (Salisbury, Wilts.); Mr K. G. Martin Leake (Asst. Librarian, Courtauld Institute of Art, 20 Portman Square, London); Mr B. S. Lloyd (Mineralogist, 14 Pall Mall, London); Mrs Constance Lodge (Librarian, Henry E. Huntington Library and Art Gallery, San Marino, California); Mr J. H. Louden (Librarian, National Library of Scotland, Edinburgh); Mr J. McKee (Wallace Collection, London); Mrs S. MacPhearson (Highgate); Mr T. J. Manby (Deputy Keeper, Doncaster Museum); Rev J. N. Marcon (Fittleworth, Sussex); Dr A. D. Morris (Eastbourne, Sussex); Mrs E. Newby (Librarian, Messrs. P. & D. Colnaghi & Co., London); Mr E. C. Pelham-Clinton (Royal Scottish Museum, Edinburgh); Dr W. D. I. Rolfe (Assistant Keeper in Geology, Hunterian Museum, Glasgow); Mrs Margaret Scott (Alexander Turnbull Library, Wellington, New Zealand); Mr E. R. Seÿd (Keeper of Zoology, Manchester Museum); Mr F. K. Simpson (Librarian, the Paul Mellon Centre, London); Brigadier Clive Simson (Fyfield, Hants.); Mr P. E. Smart (Tunbridge Wells, Kent); Professor F. A. Stafleu (Editor, *Taxon,* Utrecht, Holland); Major L. Stewart-Brown (Secretary, Jourdain Society); Mr C. K. Swann (Wheldon & Wesley, Codicote, Herts.); Miss J. M. Sweet (c/o Royal Scottish Museum, Edinburgh); Dr F. Taylor (Librarian, John Rylands Library, Manchester); Mr W. A. Taylor (Librarian, Birmingham Public Libraries); Miss D. F. Vincent (Librarian, Edward Grey Institute, Oxford); Professor Sir Ellis Waterhouse (former Director, Mellon Centre, London); Dr C. Waterston (Keeper, Department of Geology, Royal Scottish Museum, Edinburgh); and the staff of the Westminster Reference Library, St. Martins Street, London, W.C.2.

PART ONE

Introductory Articles

ENTOMOLOGICAL SALES

J. M. CHALMERS-HUNT*

For centuries, curious, rare and beautiful insects have been among the most sought-after of natural history objects. On the one hand, rarity and considerable scientific interest, and on the other the simple desire to possess these exquisite, often colourful and highly decorative treasures, have resulted in the formation of innumerable insect collections throughout the civilized world.

Entomologists usually mount their captures on pins and keep the specimens in cork-lined wooden boxes or specially constructed cabinet drawers containing camphor or some other preservative. Except for certain refinements, this system for maintaining insect collections has hardly changed since the early 18th century, and even the apparatus used in insect preparation is much the same now as it was then. The paramount importance of keeping adequate data with each specimen, however, did not receive widespread recognition until comparatively recent times, and up to a century or so ago few collectors troubled to label their captures. Nowadays, when the scientific and financial value of an insect depends as much on proper authentication as on rarity and condition, a label accompanying a specimen with particulars of at least the locality and date of its capture is of course a *sine qua non*.

A number of insect collections of great age and historic interest survive, of which perhaps the most ancient, though now only fragmentary, is that of James Petiver (1663-1718), bought on the latter's death (but not at auction) by Sir Hans Sloane (1660-1753) and now preserved in the British Museum (Natural History). Petiver's collection of lepidoptera is remarkable, not only for its great antiquity and detailed documentation, but also for the curious method of its preservation. The insects are sandwiched between layers of mica bound with gummed paper in a fashion similar to that of modern colour slides, with the data written in ink on the paper bindings of wooden frames, which have since been collected and mounted in folio volumes. This technique is somewhat similar to that employed by Henry Baker (1698-1774), whose small collection of butterflies and moths auctioned in 1775 was 'preserved between flakes of isinglass upon cards'.

Some idea of the popularity of insect sales may be gained from the fact that over 800 entomological auctions of varying extent are known to have taken place in the British Isles, of which, incidentally, the vast majority were conducted in London. The main period of vending was between 1870 and 1930, with the highest appreciable increase in the number of sales from 1880 to 1890 and their maximum frequency in 1889, 1895, 1896 and in 1905. From 1940 to the beginning of 1963 was also noteworthy as a time of exceptional insect auction activity, with seventy exclusively entomological sales being conducted then, and to my knowledge hardly a single collection being auctioned in any other branch of natural history.

The earliest entomological auctions were apparently conducted on the Con-

*1, Hardcourts Close, West Wickham, Kent.

tinent, but the first British auction in which insect specimens were offered for sale was that of John Woodward (1665-1728) in 1728. From then until the close of the 18th century there are records of some thirty insect auctions of which at least three deserve special mention.

The sale in 1786 of the museum of that great patroness of natural history, Margaret Cavendish Bentinck, Duchess of Portland (1715-1785), consisted of a remarkable assemblage of virtually all insect orders numbering many thousands of specimens scientifically arranged and totalling 344 lots. Her museum had as its curator Daniel Carl Solander (1736-1782), eminent naturalist and favourite pupil of Linnaeus. Solander compiled the sale catalogue, the markedly scientific treatment of which, unusual in auction catalogues, gave it a distinctive quality. The principal buyers were Francillon, Hunter, Drury and the naturalist dealer Humphrey. Notable among the catalogue entries are the following:

> Lot 287. Various lepidoptera, *'many of them very rare* such as *Phalaena praecox* L. [the Portland Moth] of which there are no less than 12 pair'.
> Lot 1891. Fourteen species of British coleoptera 'chiefly of the Cerambyx and Leptura genera, *amongst which is that rare and curious Insect* the Mordella Paradoxa [*Metoecus paradoxus* L.] and Attelabus Apiarius L. [*Trichodes apiarius* L.]'.
> Lot 3186. 'Three extremely curious and rare English Phalaenae, among which is that very uncommon one the Delphinii [*Periphanes delphinii* L.], or peaseblossom Moth, of Wilkes...'

The insect collection of Thomas Pattinson Yeats attracted the attention of that doyen among entomologists J. C. Fabricius (1747-1810), who described a number of species therein as new to science, including the British moth *Agonopterix yeatiana*. This collection was auctioned on 12 May 1783, but despite its obvious importance, regrettably no copy of the sale catalogue appears to have survived. Another notable 18th-century auction was that of the collection of Mr. Church, an Islington apothecary and, according to E. M. da Costa (1812), 'a great entomologist and breeder of insects'. Hardly anything is known about Church or his collection, however, and like that of Yeats no copy of the sale catalogue, as far as is known, remains.

The first half of the 19th century produced a number of very good insect sales, generally more specialised than hitherto, and with King and Lochée conducting in 1805 the first auction in Britain known to be entirely devoted to entomology. Dru Drury (1725-1803), an opulent London jeweller and the renowned author of *Illustrations of Exotic Entomology,* was described in Swainson (1840) as 'one of the most zealous and successful collectors of insects that ever prosecuted the study in this country'. His collection upon which he had spared no pains or expense consisted of over 11,000 specimens 'collected from all countries with which Britain has any intercourse....' Among those present at the sale were such famous names as J. Francillon, William Kirby (of 'Kirby and Spence'), William Swainson, old Standish the dealer, George Milne F.L.S., Alexander Macleay, A. H. Haworth and Edward Donovan, the latter's bitter adversary who also compiled the catalogue. Drury's material, very select and mostly in short series was the acme of choiceness, and to judge by some of the prices paid, the bidding must have been spirited.

> Lot 3. 'Phalaena Aprilina, [*Dichonia aprilina* L.] graminis [*Cerapteryx graminis* L.] and 22 others £7.12.

Lot 8. 'Sphinx Convoluli and 8 others' £4.1.
Lot 46. 'Papilio Priamus' £4.14.6.
Lot 95. 'Scarabaeus Goliathus var.' £12.1.6.
Lot 100. 'Cerambyx Gigas, and 2 others' £4.8.
Lot 104. 'Thirteen species of the Buprestis Genus' £8.
Lot 201. 'Mutilla bimaculata, thoracia, Scolia signata and 24 others' £3.5.

The following year King and Lochée conducted a very different type of sale, that of the Leverian Museum. Sir Ashton Lever (1729-1788), a wealthy squire of Alkrington Hall near Manchester, had a passion for collecting all kinds of natural objects. His museum, first housed near Manchester then in London, was acclaimed the first in Europe, and for his efforts in creating it he was knighted in 1778. The museum's upkeep was too costly, however, and though offered to the government for a sum far below its estimated value of £53,000, this was refused on the prejudiced advice of Sir Joseph Banks, who hated both Lever and his museum. In 1786, it was sold in a lottery of 8000 tickets at a guinea each but the new owner was eventually forced to dispose of it for financial reasons and the museum was finally auctioned in 1806. The sale of this vast miscellany included some 240 insect lots among which were many interesting items, for instance:

Lot 4051. 'Cetonia hamata' Donovan £3.3.
Lot 4052. 'Phasma gigas, in the winged state, *very rare*' Macleay £4.
Lot 85 (supplementary) 'Scarabaeus Hercules, male and female, American - the latter very rare' P. Jackson £3.5s.
Lot 89 (supplementary) 'Phasma dilatatum, a large and fine species. This specimen is presumed unique' Macleay £11.10s.

During the next fifteen years four more important collections were auctioned, those of Simon Wilkin (1790-1862), Edward Donovan (1768-1837), Thomas Marsham (c.1748-1819) and, in 1818, that of John Francillon (1744-1816). Donovan, a natural history painter and the author of several finely illustrated works among them *The Natural History of British Insects,* must have had one of the most extensive entomological collections of his day. In 1817, he advertised his museum, the London Museum and Institute of Natural History, for sale which, according to Parkin (1911:5) and others, was auctioned in 1818. But the details are curiously lacking, for although Parkin (*op.cit*) mentions consulting a copy of the sale catalogue which Alfred Newton had bequeathed to the Zoological Museum, Cambridge, it has not yet been traced and there appears to be no other copy extant.*

The Wilkin collection, particularly rich in coleoptera, consisted of some 10,000 insects of all orders including those of Dr. Beckwith and Sir Joseph Hooker. Neither of the two surviving copies of the Wilkin sale catalogue is marked, but the anonymous annotator in Sherborn (n.d.) indicates that the insects finally went to the British Museum (Natural History). Francillon, a London doctor, had it is said 'the most magnificent cabinet of insects that has ever been brought to sale in this country'. Be that as it may, its chief importance was that it contained many Fabrician as well as numerous Donovan types described by that author in his *Epitome of the Insects of New Holland* and elsewhere. Alexander Macleay bought many of the lots, and according to Horn and Kahle (1935:80) part of the

---

*After much enquiry I have established that the Donovan sale catalogue still survives and am able to give particulars of it in the *Register*.

Francillon insect collection eventually reached the British Museum (Natural History) and part the Hope Department, Oxford. Finally, there was the Marsham sale in which the important cabinet of British coleoptera described in his *Entomologia Britannica* was offered in one lot and which J. F. Stephens bought. Marsham was a founder member of the Linnean Society, and in 1798, its Secretary.

The year 1834 is a memorable one in the annals of entomological history as that in which J. C. Stevens offered to the public what was probably the most outstanding insect collection ever to be auctioned. Adrian Hardy Haworth (1767-1833), foremost among the entomological cognoscenti of the early 19th century and a pioneer of modern insect study in this country, was the author of two signal works - both of which are now exceedingly rare - *Prodromus Lepidopterorum Britannicorum* (1802), and the classic *Lepidoptera Britannica* (1803-1828). The latter, the first book to treat of the British lepidoptera on a general and scientific scale, has been described as 'a monograph the most complete, most learned, and most useful, ever published on the Entomology of Britain'. The sterling qualities of Haworth were firstly, that he described from Nature and, secondly, that he described well. His collection of nearly 40,000 specimens of virtually all orders and of which more than half were lepidoptera, contained a large proportion of the species described in the *Lepidoptera Britannica,* many of which are now in the British Museum (Natural History), while others are in the Hope Department, Oxford, and some in the Yorkshire Museum, York. They bear his original trapezoidal-shaped labels, and the characteristic small square blue tickets with which his British insects were marked. It has been said that every naturalist of any note was present at the Haworth sale, among them Edward Newman, John Curtis, Swainson, Chant and Francis Walker. One of the most interesting items was Lot 898, the first British specimen of the curious Strepsipteron *Stylops haworthii* Stephens for which Haworth once refused the offer of 25 guineas, but which on this occasion fetched only £1.6s. J. O. Westwood (later Hope Professor of Entomology at the University of Oxford) had compiled the catalogue and the sale, which lasted eleven days, realised £552.12s, a sum far less than was expected, as is born out by an inscription which Adam Smith wrote 5 October 1858 on the title-page of the Gowing-Scopes copy of the catalogue: 'The results of the sale much disappointed the [Haworth] family as I have heard at the time Exotic Insects were not bought and so fetched but little'. In March 1834, Sotheby sold Haworth's magnificent library, for years the only one to be auctioned in any way comparable with the great entomological library sales of Children (1840), Stainton (1899), Mason (1904) and Chapman (1922).

The general collection of George Milne (c.1765-1838), sold in 1839, consisted of 17,000 specimens of which much of the foreign material was bought at the Drury, Francillon and Marsham sales. The only surviving copy of the catalogue of this important collection is unmarked, but according to Horn and Kahle (1936:177) the British Museum (Natural History) acquired a portion of the lepidoptera and coleoptera. The sale of the insects of Sir Patrick Walker (1772-1838) in the same year is interesting on account of the number of curiosities it contained, among them Lot 174 which included the historic specimen of 'Spilosoma Walkerii' [*S. luteum* Hufnagel, var.]. This strikingly unusual moth was first depicted and described in Curtis, *British Entomology* (1825), as a species new to science, and was again figured as recently as 1974 in the latest edition of South, *Moths of the British Isles*. Two other important insect collections sold about the same time as Milne's were those of J. G. Children (1777-1852) and William

Swainson (1789-1855). Children was on the British Museum (Natural History) staff and the first Keeper of the zoological collections, but he possessed a private general insect collection which was auctioned in 1840. This consisted of 'upwards of 37,000 specimens' many of which the British Museum (Natural History) bought and many of which Francis Walker and others described as new species. Swainson, the author of the handsome *Zoological Illustrations* and other notable works, had a collection of virtually all orders of insects, though particularly rich in hymenoptera and diptera and containing a number of types.

No catalogue survives of the sale in 1844 of the collection of the Entomological Club, but a minute-book referred to by South (1899) reveals among other things that it embodied a considerable quantity of foreign insects. A little-known sale held in 1846, of the insects of Dr. Warburton, is of interest as Sherborn (1940:139) states that he 'had some Haworth types'. Of their fate though we have no knowledge.

The Victorians may have been dilettanti but there is no doubt they advanced scientific knowledge enormously and by no means least the study of entomology. During this period, there appeared a spate of specialised literature including a number of magazines and key-works for identification that were to have a profound influence on the development of entomological research for many years to come. At the same time, interest in insects and particularly insect collecting underwent a spectacular rise, resulting in the formation of a great many collections numbers of which, including some of considerable significance, were later to pass through the sale rooms.

Of no scientific interest at all, but exemplifying 19th-century taste, were decorative butterfly arrangements of intricate design. These, partly in the form of a date if commemorating some special historical event, were particularly popular in Victorian times. Exhibited within a wall frame or sometimes beneath a glass dome to rest on a table, such frivolities were not unknown at an earlier age, with two examples passing through the saleroom in 1794. At the Jackson sale on 2 May that year Lot 69 consisted of a 'Small collection of butterflies collected by Mr. I. Jackson and forming the initials of his name'. And at the sale of an insect collection of 'A Gentleman left off Collecting' on 2 June, Lot 84 was described as being arranged 'in a large frame, glazed with plate glass.....representing a piece of architecture.'

In 1850, was auctioned the library of one of the most illustrious of English entomologists, the Rev. William Kirby (1759-1850), a hymenopterist *par excellence* and the author of *Monographia Apum Angliae* (1802) and co-author with William Spence of the famed *An Introduction to Entomology* (1815-1826). Kirby kept a *Journal* from which were published some fascinating extracts. This unique document seems to have disappeared, however, and a perusal of the sale catalogue shows that it was not listed among the items therein.

The following two sales are mainly of interest as relating to the early history of the Royal Entomological Society. Owing to the state of chaos into which the Society's collections had fallen, a special committee recommended their disposal and in 1858 Stevens auctioned a portion of the exotics and a further portion of these in 1863, together with the British insects. As the Society had been the beneficiary of numerous collections such action, as may be imagined, gave rise to severe criticism and dissension among the members, and following the committee's recommendations an acrimonious correspondence ensued in the pages of the *Entomologist's Weekly Intelligencer* between one signing himself 'M.E.S.' and one 'A Member of Council'.

In 1863, there were auctioned John Walton's (1784-1863) fine collection, so rich in coleoptera especially British Curculionidae; in 1865 and 1867, the insects and library of the Rev. Hamlet Clark (1823-1867); and in 1868 the collection of British insects of Thomas Desvignes (1812-1868), including W. E. Shuckard's (1803-1868) celebrated collection of hymenoptera.

The last quarter of the 19th century was characterised by the highest level ever reached in the number of insect auctions conducted. Many of these were of collections of British lepidoptera that had been augmented by postal exchange, a system of barter originating about 1850 which operated for at least half a century and reached its zenith in the 1880's. The final decades of the Victorian era also witnessed some remarkable changes: for instance, prices of insects especially of certain lepidoptera rose to unprecedented heights, with the increasing difference in value of a British insect compared with that of a foreign example of the same kind becoming most marked. This is shown by the huge prices paid at auction for rare immigrant species relatively common abroad; a Continental specimen of *Argynnis lathonia* L. (Queen of Spain Fritillary) or *Pontia daplidice* L. (Bath White) for example, though listed in dealers' catalogues at only a few pence each, might if British fetch as many pounds at auction. Inevitably this state of affairs gave rise to imported specimens being labelled as genuinely English by the unscrupulous, and in due course being bought by the gullible for sums far exceeding their original cost. Indeed, Edwin Birchall, writing in 1877, observed that whatever may have formerly been the case it had by then become impossible to make a private British collection of lepidoptera unless the collector restricted himself to specimens of his own capture. Interest in varieties of British lepidoptera also grew enormously at this time, and striking aberrations became increasingly sought-after, with prices at auction reaching exceptional levels, especially among the butterflies.

The many insect auctions of this prolific period included some of considerable importance. Edwin Brown (1819-1876) of Burton-on-Trent, had an extensive knowledge of all departments of Natural History and his ample collections, housed in a large room adjoining his residence, contained entomological as well as other groups. These were sold in 1877 and of the 915 insect lots, 700 were of coleoptera. The following year much of the material of Andrew Murray (1812-1878) was dispersed, including his types of Old Calabar coleoptera which Janson bought for the British Museum (Natural History).

Murray was the author of many notes and papers on the coleoptera, among them a *Catalogue of the Coleoptera of Scotland* (1852); but probably his most important contribution, a *Monograph of the Nitidulidae* (1864), was never completed. In 1880, the library and a portion of the collection of Frederick Smith (1805-1879) was sold. Smith was one of the greatest hymenopterists this country has ever produced, a member of the staff of the British Museum (Natural History) and the author of some 150 entomological papers many of high importance, as well as several major works including various Catalogues of British and World hymenoptera issued by the British Museum.

Between 1880 and the turn of the century, so many important insect sales took place most of which were of British lepidoptera, that it is possible only to mention a few of the more notable collections briefly. In 1882, Stevens auctioned the collection of John Sang (1828-1887) of Darlington, an excellent microlepidopterist and entomological artist, and the discoverer as new to science of *Syncopacina sangiella* Stainton and *Monochroa tetragonella* Stainton; and in the following year the collection of all insect orders of Benjamin Cooke (1816-1883) of

Southport, the first to discover the very local moth *Lycia zonaria* D. & S. in Britain, as well as numerous other additions to our fauna. Dr. Philip Henry Harper (1822-1883) had amassed a large assemblage of the choicest varieties of British butterflies and moths, for which in 1884 the prices paid were probably higher than had ever before been given for lepidoptera. This was especially so of the Geometridae, among which a series of *Abraxas grossulariata* L. (Magpie Moth) fetched over £100. It was at this sale too that the then unique British example of *Lycia lapponaria* Boisduval changed hands at £13.13s, the price of which would probably have risen far more had not one of the competitors understood that it had been knocked down to him, and so ceased bidding.

One of the richest collections to be auctioned during this period was that of Howard Vaughan (1846-1892), in 1890. Vaughan, a superlative lepidopterist, particularly in the field, and the discoverer of *Rotruda saxicola* Vaughan as new to science, possessed many highly desirable species especially among the microlepidoptera, including *Catoptria verellus* Zincken, *Chrysocrambus linetella* Fabricius (*cassentiniellus* H.-S.), *C. craterella* Scopoli (*rorella* L.) and *Pammene obscurana* H.-S. In 1893 and 1894 the extensive collection of British lepidoptera formed by the Rev. Henry Burney (1813-1893) was brought under the hammer. The principal feature of this sale was the great number of rare species it contained, many of them in more or less long series, as for instance 31 specimens of the superb but extinct (since 1848) indigenous British butterfly *Lycaena dispar* Haworth (Large Copper). Other remarkable insects were the type of *Luperina nickerlii gueneei* Doubleday; also, seven specimens of *Paracorsia repandalis* D.& S. a species Burney discovered and bred about 1876, and which so far as is known has not occurred in Britain since. The genuinness of much of his other material was suspect, however, for though of unquestionable integrity himself, Burney tended to buy indiscriminately and on the whole many of his rarities realised relatively low prices.

William Machin (1822-1894), a practical entomologist of exceptional expertise and most successful rearer of microlepidoptera, had from the age of fifteen made a collection composed entirely of British species. Unlike some well-known collections of this period, Machin's was renowned not only for its perfection but also for the reliability of the specimens it contained, many being of his own taking. Not surprisingly, therefore, at this sale in 1895 prices were very stiff and in many instances reached high-water mark. It was to Machin that we owe *Coleophora vibicigerella* Zeller and *C.machinella* Bradley as British species. The following year the collection of a solicitor, Charles Adolphus Briggs (1849-1916) was sold. An eclectic buyer of British lepidoptera with an eye for historical authenticity and the accompaniment of accurate data with his purchases, Briggs had a particularly interesting assemblage, in which his own captures were largely supplemented by those from the Farren, Vaughan, Machin, Tugwell, Burney, Robson, Weir, Mitford and Prest sales. Poorly arranged lotting, however, emphasised the fact that many varieties were conspicuously similar, in consequence of which prices for these on the whole ranged low, though by contrast those of some of his other insects were exceptionally steep, with one Large Copper reaching the then record figure of eight guineas.

In 1897, the British lepidoptera of J. B. Hodgkinson (1823-1897) was auctioned. He was an indefatiguable field-worker, especially among the smaller moths, and had as a lad lost an eye (blown out when shooting) but he could to the last see as much with one eye as most people can with two. Despite his disability, Hodgkinson's discoveries included a species new to science, *Coleophora adjunctella* Hodgkinson, with a wing span of only eight millimetres! The sale of George Elisha's lepidoptera the following year was in the nature of a tragedy, for the condition of the specimens and excellence of setting this collection was probably unsurpassed, but owing to the absence of localities the prices realised were among the lowest ever at auction.

Samuel Stevens (1817-1899) brother of J. C. Stevens and uncle of Henry Stevens, conducted the auction business in King Street for some years. An ardent field-collector and discriminating buyer in the rooms, his huge butterfly and moth collection of sixty years of search and care was wonderful. Entomologists from all parts of the country flocked in 1900 to this historic sale, not only because of its connexion with the auction family, but also for the amplitude of rarities, extraordinary varieties and numerous Haworthian types offered.

Other notable collections auctioned about this time were those of the Rev. Thomas Blackburn (1844-1912), in 1870; W. W. Saunders (1809-1879), in 1874, much of whose material was acquired by the British Museum (Natural History) and Hope Department, Oxford; George Wailes (1802-1882), of Newcastle-upon-Tyne, in 1884, a number of whose captures from 1828 onwards were recorded in J. F. Stephens, *Illustrations of British Entomology;* Sir Sidney Smith Saunders (1809-1884), in 1884, who was a cousin of W. W. Saunders (*q.v.*) and an authority on the hymenoptera, strepsiptera and certain other orders; Major F. J. S. Parry (1810-1885), in 1885, a coleopterist specialising in the Lucanidae and Cetoniidae; Walter Battershell Gill (1823-1900), in 1886; William Warren (1839-1914), in 1888; A. F. Sheppard (M.E.S., 1851-1875), in 1889; J. S. Baly (1816-1890), of Warwick, in 1890, a coleopterist specialising in the Phytophaga; Ferdinand Grut (1820-1891), in 1891, a coleopterist and sometime Librarian of the Entomological Society of London, whose collection contained exotic Cicindelidae, Carabidae, Dytiscidae and Gyritidae; Rev. F. O. Morris (1810-1893), in 1893, so well-known for his popular illustrated books on diverse natural history subjects, and the author of several anti-Darwin pamphlets; J. Jenner Weir (1822-1894), in 1894; J. E. Robson (1833-1907), in 1895, author of the splendid *Lepidoptera of Northumberland, Durham and Newcastle-upon-Tyne;* William Farren (1836-1887), in 1895, a Cambridge lepidopterist whose magnificent series of *Cryphia muralis impar* Warren included those described by J. W. Tutt; W. H. Tugwell (1831-1895), in 1895; and finally, the sale of Lord Dormer's (1830-1900) exotic Cicindelidae in 1901.

Insect auctions conducted during the first decade of the 20th century maintained on average a fairly high level of frequency, but underwent a very marked decline during the First World War, to reach in 1915 the lowest number on record for forty years. Much saleroom activity followed the end of the war with a maximum of eleven entomological auctions in 1919, which annual total has not been reached since. In 1940, not a single insect auction was held nor in that year one in any other branch of natural history to my knowledge, the first time this had happened since 1772. Curiously enough this hiatus was not maintained in the years immediately following, and during the Second World War insect auctions actually showed an appreciable increase in their numbers. However, since 1960, when again no insects were auctioned, there have been on average fewer entomological sales and none of any collection of real importance.

Up until the First War, there were among those auctioned, seven insect collections deserving special mention, of which all except one were of lepidoptera: those of Philip Brookes Mason (1842-1903); Charles Golding Barrett (1836-1904); Alfred Beaumont (1832-1905); Rev. Henry Charles Lang (decd. 1909); Thomas Maddison (decd. 1908); John Adolphus Clark (1842-1908); and, in 1911 and 1912, probably the most significant of them, that of James William Tutt (1858-1911).

Mason, a Burton-on-Trent doctor of medicine and an entomological and botanical savant, was the possessor of a fine natural history museum containing probably, the most nearly complete private collection of British lepidoptera ever formed. Auctioned in three parts in 1905, Mason's collection incorporated that of Edwin Shepherd (Secretary Ent. Soc., 1855-1866) as well as many specimens bought from the chief collections auctioned over the past twenty-five years. One of the principal features of the Mason sale was the unusually large number of historic examples it embodied, including some A. H. Haworth, John Curtis and J. F. Stephens type material. Amongst this *embarras de richesse* was the unique Psychid, *Solenobia douglasii* Stainton which, offered with 44 other insects in Lot 155 on 28.xi.1905. was sold for the extraordinarily low figure of 11s. Luckily however, the purchaser was E. R. Bankes, whose collection, together with the priceless *douglasii,* was later presented to the Nation. The following year Beaumont's valuable hymenoptera were auctioned, which apparently A. J. Chitty bought, and so were destined to enrich the collections at the Hope Department, Oxford.

In 1906 and 1907, Stevens auctioned the collection of lepidoptera of C. G. Barrett, the celebrated author of *The Lepidoptera of the British Islands,* now some seventy years old but still the standard work on the subject. Among Barrett's many gifts were his untiring energy and enthusiasm, and we find in 1856 a note from him relating how he saw from the train two Clouded Yellows (*Colias croceus* Geoffroy) flying on the railway embankment near Forest Hill, and got out at the next station and went back and caught them. Among the more interesting things at this sale were the type specimens of *Mythimna favicolor* Barrett and its beautiful reddish form. Fortunately, as with many other insect types at auction, these were purchased by Lord Rothschild, this country's foremost benefactor of entomological specimens, and whose vast collections at Tring were later incorporated with those of the British Museum (Natural History). Lang, the author of the well-known *Butterflies of Europe* (1884) in two quarto volumes, had a particularly fine collection of Palaearctic butterflies. But, although the specimens were in splendid condition and nearly all the rarest species were represented, very low prices were realised at this sale in 1907.

The value of a really good collection of British lepidoptera was well illustrated by the Maddison sale in 1909. At this, the prices on the whole were relatively high, though perhaps the most remarkable, and possibly valuable specimen scientifically, was a Garden Tiger (*Arctia caja* L.) with five wings from the Gregson collection and figured by Mosley which, together with various others in this lot, went for only 11s. In 1909 and 1910, was auctioned the collection of the London naturalist J. A. Clark, one of the founders of the Haggerstone Entomological Society (later to become the City of London Entomological and Natural History Society), whose series of varieties of *Arctia caja* L. and *Acleris cristana* D. & S. were among the principal high-lights of this sale.

Many of the hundreds of descriptions of new races and aberrations in J. W. Tutt's monumental *The British Noctuae and their Varieties* (1891-1892) and his *magnum opus, A Natural History of the British Lepidoptera* (1899-1914), were based on specimens in his own collection, few if any of which were ever figured. That this unique assemblage should have been dispersed is one of the most regrettable facts associated with entomological auctioning, and a disaster for future workers engaged in the study of the lepidoptera and their variation. Ironically too, were the extremely low prices realised at the Tutt sales, with often two or more lots having to be combined to find a buyer. The whole of this collection should have (and it undoubtedly could have) been secured for the Nation, and one fails to understand why this was never done.

The final sixty years of the period under review saw the auctioning of comparatively few collections of other orders besides lepidoptera, and only three of these it seems were of special note. Those of the British coleoptera of Frederick Charlstrom Adams (decd. 1920), in 1919; the parasitic hymenoptera and coleoptera of Thomas Richard Billups (decd. 1919), in 1920; and in 1926, some of the insects of Thomas Ingall (M.E.S., 1849-1856), which St Bartholomew's Hospital sold, including thirty Haworthian types that Rosenberg bought for the trifling sum of five shillings.

Before proceeding further, reference must be made to the Rev. Gilbert Henry Raynor (1854-1929), who had twelve insect sales (more than anyone else) spread over a period of forty years. His sales were mostly of varieties of the Magpie Moth (*Abraxas grossulariata* L.), and it was the remarkable results he achieved from breeding this species for which he is best remembered. Shortly before his death, Raynor is reputed to have deliberately destroyed one or more of his best strains of *grossulariata,* in order it has been said that the existing specimens should not lose their value. He appears moreover, to have been of a somewhat reticent disposition, and one suspects that much of his knowledge of the genetics of this species and breeding technique accompanied him to the grave.

The great interest in collecting varieties of British butterflies that came about during late Victorian times increased if anything from around 1920, with prices of the rarer and more striking forms reaching an all-time high in the 1930's and 1940's. On the other hand, comparatively little interest was shown by the great majority of collectors in the factors that cause these curious and often exceedingly beautiful abnormalities. With such a rich store of varietal material, therefore, as has now accumulated and the splendid example set by the researches of the late Dr. E. A. Cockayne, here is indeed a golden opportunity for someone to investigate this fascinating but neglected field.

It was in 1919 and 1920 that Sydney Webb's (1837-1919) British lepidoptera were offered. Generally considered to be the finest collection of these ever put up for sale, it had magnificent series of varieties and aberrations (many of which have been figured and described in the literature), the result of his own indefatiguable labours coupled with those of Frederick Bond (1811-1889) and Charles Stuart Gregson (1817-1899), whose entire collections at their deaths, Webb embodied in his own. These collections also included, among others, specimens purchased at the Briggs, Gill, Mason, Harper, Hopley, S. Stevens and Vaughan auctions. Among Webb's great rarities were the following: *Ochropleura flammatra* D. & S., *Meganephria bimaculosa* L., *Lithophane lambda* Fabricius, *Lacanobia blenna* Hübner, *Chrysocrambus linetella* Fabricius, *Glyphipteryx lathamella* Fletcher and a host of others too numerous to mention, to say nothing of a magnificent series of 1290 *Acleris cristana* D. & S. including the Clark, Mason and Webb types.

Thereafter a dozen other collections were auctioned that were almost as rich in varieties. Those of Alfred Bridges Farn (1841-1921), Arthur Horne (1864-1922), Sir Vauncey Harpur Crewe (1846-1924), Benjamin Hill Crabtree, Herbert Massey, Frederick Janson Hanbury (1851-1938), Baron Bouck, Percy May Bright (1863-1941), William George Sheldon (1859-1943), Sir Beckwith Whitehouse (1882-1943), Benaiah Whitley Adkin (1865-1948) and the Rev John Marcon (b.1903). The Bright sales in 1938, 1941 and 1942 probably constituted the richest assemblage of British butterfly varieties ever to have been auctioned. One reviewer said Bright 'possessed a collection the like of which will hardly be equalled much less excelled, for many years to come'. Among his most spectacular items were the all white and all black varieties of the Marbled White (*Melanargia galathea* L.) - ab. *totaalba* Chalmers-Hunt and ab. *lugens* Oberthür, which were resold in 1943 together as one lot at the Whitehouse sale for £110, the highest figure ever obtained at auction for two British insects.

A revival of interest in Continental and exotic lepidoptera resulted in fairly high prices being paid at Dr. Michaud's sale in 1961, and at the collection belonging to Canon George Watkinson auctioned the following year. But the ultimate at auction, at least in this country, was reached in 1970, when among Dr. Lewis G. Rosa's exotics, Laithwaite paid £130 at Sotheby's for a pair of the magnificent New Guinea bird-wing butterfly *Ornithoptera joiceyi* Noakes & Talbot.

But what of the future of entomology at auction? Recent sales show that mounted specimens especially of rare and showy butterflies, are again much in demand with prices higher than ever before, though the signs are that those who can afford to buy nowadays do so more for an insect's visual appeal than for any scientific interest. Indeed, it is as likely as not that today's purchaser knows little or nothing about entomology, nor has any wish to do so. Are we, therefore, approaching a period when the average entomologist can no longer afford to buy in the salerooms because some rich interior decorator or designer will outbid him? Only time will tell, but the indications point that way and, though legal policy for conservation safeguards is already in force in some countries, it would be sad if vanity or fashion in adornment were to result in the extinction of some of the world's rarest, scientifically interesting and most beautiful butterflies.

REFERENCES

Allingham, E. G. 1924. *A Romance of the Rostrum*. Witherby, London.

Costa, E. M. Da. 1812. Notes and Anecdotes of Literati, Collectors, &c., collected between 1747 and 1788. *Gent. Mag.* (1), **1812**: 205-207, 513-516.

Gilbert, P. in press. *Compendium of Biographical Literature on Deceased Entomologists*.

Horn, W. & Kahle, I. 1935-37, 1961. *Uber entomologische Sammlungen, Entomologen & Entomo-Museologie*, with *Supplement* by Sachtleben, H. Berlin.

Kloet, G. S. & Hincks, W. D. 1972. *A Check List of British Insects* (2nd Edition Part 2 completely revised by Bradley, J. D., Fletcher, D. S. & Whalley, P. E. S.). Royal Entomological Society, London.

Neave, S. A. & Griffin, F. J. 1933. *The History of the Entomological Society of London, 1833-1933*.

Parkin, T. 1911. *The Great Auk: A Record of Sales of Birds and Eggs by Public Auction in Great Britain, 1806-1910*. Hastings and East Sussex Naturalist, Supplement to Part 6, Vol. 1.

Sherborn, C. D. 1940. *Where is the – Collection?*

Sherborn, C. D. N.d. *A MS. Supplement* (in an unidentified hand) to, and incorporated in, a copy of *Where is the – Collection?*

Smith, W. J. 1960. A Museum for a Guinea. *Country Life*, **127**:494-495.

South, R. 1899. The Entomological Club. *Entomologist*, **32**:160-164, 224-226.

Stearn, W. T. 1965. *A. H. Haworth: Complete Works on Succulent Plants*, Vol. 1: *Biographical and Bibliographical*.

Swainson, W. 1840. *Biography of Zoologists, and Notices of their Work.* Lardner's Cabinet Cyclopaedia.

Wilkinson, R. S. 1966. English Entomological Methods in the Seventeenth and Eighteenth Centuries. *Entomologist's Rec. J. Var.* **78**:143 *et seq.*

In addition to the foregoing I have freely used information from obituaries of entomologists and reviews of sales in the following periodicals: *Entomologist's Record and Journal of Variation, Entomologist,* and *Entomologist's Monthly Magazine.*

ZOOLOGICAL SPECIMENS
OTHER THAN INSECTS, MOLLUSCS, AND BIRDS

ALWYNE WHEELER*

Collections of molluscan shells, many insects, birds and their eggs have a considerable aesthetic appeal of their own, in addition to their intrinsic value. The object collected is usually individually small, frequently brightly coloured, preserves well in a dried state and is thus well suited to the collector's needs. Not surprisingly, with herbarium specimens and mineral collections, such objects have tended to dominate natural history sales in the past. Other zoological groups rarely lend themselves to collecting in the same way. Collections of mammals, reptiles, fishes, and invertebrates other than insects and molluscs are rarely as satisfying aesthetically as those of the more popular groups. For example, comparison of a drawer filled with butterflies or a case of minerals with a collection of dried crabs or bottled snakes will rarely be to the latters' advantage. Animals such as fishes, amphibians, reptiles, crustaceans, and to some extent mammals, do not have the intrinsic appeal necessary to encourage the collection of their remains and, moreover, their preparation and display is much more difficult than for the more commonly collected groups. It is not therefore surprising that collections of these animal groups are relatively infrequent in the saleroom.

There are, however, two exceptions to this. Collections made for serious study, particularly those which contain specimens cited in the publications of early naturalists, have an historical and scientific value which far outweighs their aesthetic or intrinsic appeal. An example of a collection of this kind is the sale of John Ellis's (1710-1776) corals, corallines, and sponges in 1791, many of them named by Daniel Solander and which formed the basis of two of Ellis's publications, *An essay towards a natural history of the Corallines ...* (1755) and *The natural history of ... zoophytes* (1786). Another example is the collection of fish skins which appeared on the London market in 1853, and which later investigation by J. E. Gray proved to be the major part of the fishes formerly in the collection of Laurens Gronovius (1730-1777), the author of the *Museum Ichthyologium* (1754-56) and the *Zoophylacium Gronovianum* (1763-81). While no one could regard this collection of half skins of fishes mounted on paper as an object of beauty its importance to systematic ichthyology was profound, for it contained numerous type specimens of Linnaean names of fishes.

The second exception to the general rule concerns collections of trophies of the chase. Birds and fishes were frequently preserved stuffed and cased, but game mammals were more usually saved for display as mounted heads, although the smaller species may be kept in glass cases. Collections of this kind are more subject to economic and social changes than other kinds of natural history collection and for reasons to be discussed later are now rarely brought together.

Among the general collections sold in the 19th century several contained a

*Department of Zoology, British Museum (Natural History), Cromwell Road, London SW7.

mixture of zoological specimens. Thus, the Leverian Museum sale of May-July 1806 contained numerous lots of mammals, reptiles, and fishes; among the latter at least were several specimens known to have been collected on Cook's voyages. One of the main buyers at the Leverian sale was Edward Donovan (1768-1837) whose London Museum and Institute of Natural History was well known. It contained a remarkable collection of Recent British quadrupeds, reptiles, fishes, crustaceans, and zoophytes, besides fossils. According to the prospectus in *The Gentleman's Magazine,* May 1817, the whole was to be offered for sale early in 1818. In this sale presumably the type specimens of fishes described by Donovan in his *The natural history of British fishes* (1802-1808), were amongst the 130 fish species said to have been included. There were two series, those in cases and a separate series in spirits; both contained a number of species of very restricted distribution such as the charrs of Windermere and North Wales, and the gwyniad of Llyn Tegid, both fishes which were very poorly known at that time. The collections amassed by William Bullock, and sold under several names (London Museum of Natural History in May and June 1819; Mexican Exhibition in 1825; and Zoological Museum, Egyptian Hall, Piccadilly, in February-March 1828) contained specimens of mammals, reptiles, and fishes, in addition to other zoological material. Another museum collection, J. Heaviside's Museum (of George Street, Hanover Square, London) sold in 1829, contained quadrupeds and fishes, as well as other material.

Contemporaneous with Heaviside's collection was the sale of the famous Anatomical and Zoological Museum of Joshua Brookes, F.R.S., F.L.S., F.Z.S. (1761-1833), between July 1828 and March 1830. Joshua Brookes, was a famous surgeon and medical teacher of his day, who invented a method of preserving objects for his lectures and for public dissections so that they retained their colour and prevented further decomposition. His museum is said to have cost £30,000 to amass, much of the comparative anatomical material coming from his brother's menagerie in Exeter Change. Brookes's collection was general, including quadrupeds, reptiles, and fishes, but it also contained such esoteric items as Entozoa and thirty specimens of bats. A few specimens of mammals from Brookes's collection were later identifiable in the British Museum collection.

Many sales, however, contained elements of a more specialised nature, often as rather surprising additions to the material for which they are well known. For example, the collection made by the Duchess of Portland (sold in 1786), famous for its molluscan shells, also contained corals and mammalian specimens, while the associated collection of the Rev John Lightfoot, chaplain to the Dowager Duchess, included reptiles and fishes.

Other sales which included mammals or quadrupeds in their catalogues were those of Thomas Pattinson Yeats (1782) probably from Cayenne, P. Dick (1821) of Sloane Street, London, of which Lot 47 comprised a 'monster kitten with 8 legs, ... duck with 4 legs ... dog with 2 heads', and in the same year the Royal Museum, of 28 Leadenhall Street, London. In 1838 the Museum of the Cape of Good Hope Association for Exploring Central Africa was sold at Stevens' Rooms and included both quadruped and reptile specimens. This collection had been exhibited in the Egyptian Hall in 1837 and included many specimens of mammals and reptiles collected and described by Dr Andrew Smith. It was sold by auction to raise funds for further exploration.

Part of the collections made by Surgeon-Major T. C. Jerdon (1811-72), author of *The mammals of India* (1867), was sold in 1850, although he had earlier presented (1846) a large collection of Indian mammals to the British Museum. In

1863 part of Paul du Chaillu's collections of mammals and birds were sold. Du Chaillu was well known for his explorations in West Africa, detailed in his *Explorations and adventures in equatorial Africa; with accounts of ... the gorilla ... and other animals* (1861) in which he recounted some horrific stories of the behaviour of the gorilla. Later collections of mammalian remains tended to be of mounted heads and horns, trophies of the chase rather than collections of scientific or historical importance.

Sales in which reptiles and fishes feature were considerably more frequent in the 18th and early 19th centuries than later. Collections made by John Woodward, F.R.S. (1665-1728), the egregious polymath and Professor of Physic at Gresham College were sold in November 1728 and included both animal groups. Four different 'frog-fishes' from Surinam made up Lot 20 of the sale of the collection of Colonel Bosc de la Calmette of Maastricht, Holland, with other fishes and reptiles, in 1778. According to the note in the catalogue these had been described by Merian in the *Philosophical Transactions*. This is clearly a mistaken reference to the note by G. Edwards (*Phil. Trans. Roy. Soc. 51* (2), p. 653; 1760) in which he describes an amphibian and tadpoles from Surinam, sent by way of Barbadoes to John Fothergill, and discusses and reproduces Maria Merian's figures of amphibians and their larvae in her *Metamorphosis Insectorum Surinamensis*.... The interest of these specimens is considerable if they were truly the specimens illustrated by Maria Merian. As the text of her account of these animals shows, they were figures prepared from specimens in Albertus Seba's vast collection. It is possible, therefore, that the de la Calmette sale may have contained other Seba material, a possibility which has not previously been considered, despite the careful study by M. Boeseman of the dispersal of Seba's collections (*Zool. Meded., Leiden 44* (13), p. 176-206; 1970).

T. P. Yeats's reptiles, sold with the quadrupeds in 1782, may also have come from north-eastern South America. Reptiles from the collection of William Swainson were included in a sale in July 1840, and in the same year a large collection of fishes, made by Dr. Janvier in the seas around Mauritius, was sold (many, if not all, of which are now in the British Museum (Natural History) collection). In 1853 the Gronovius fish skins appeared on the London market, mysteriously imported from France with non natural history objects. Possibly, like parts of the Seba collections, they had been removed from Holland to France as the spoils of war in the late 18th or early 19th century, for there is no record of their history after the sale of the huge Gronovius collections in Leiden in 1778. Another important collection of fish specimens which were again mostly half-skins, dried and mounted on card, was that of William Yarrell (1784-1856) which was purchased by the British Museum in 1856. In 1863, part of the Linnaean Society's general collection of specimens was sold by J. C. Stevens. This included the remains (head and fins only were preserved) of J. W. Bennett's fishes from Ceylon described in his *A selection from the most remarkable and interesting fishes found on the coast of Ceylon...* (1830), and specimens collected by Mungo Park in Sumatra in 1791-92. Both collections were purchased for the British Museum. Unfortunately, only part of Mungo Park's collection can now be found, although it includes a number of type specimens of species described by Park, but Bennett's collection of fishes' fins has disappeared completely.

Compared with collections of molluscan shells which seem always to have been popular, other marine invertebrates are generally ignored by collectors. Probably the most important sale to have taken place within this category is the collection of John Ellis's corals and zoophytes already mentioned. Crustacean specimens featured in the Delafons sale of 1834, and with zoophytes in addition at the James Burton Jr. sale in 1836. In 1880 the collection of crustaceans belonging to Professor Thomas Bell, F.R.S. (1792-1880) was sold; Bell was the author of *A history of British stalk-eyed Crustacea* (1853). A year later a collection of recent crustacea belonging to John Bell was sold. Was this the residue of the Thomas Bell collection? In 1893 the collections of the Yorkshire naturalist, the Rev. F. O. Morris (1810-1893) were sold, including starfishes and foreign crustaceans. Thereafter such collections very rarely reached the saleroom. One of the few exceptions was the sponge collections of W. Saville Kent in February 1909; Kent was the author of *A manual of the Infusoria...and the organisation and affinities of the sponges* (1880-1882), as well as a detailed book on *The Great Barrier Reef of Australia* (1893).

Anonymous sales have frequently contained specimens of all the animal groups. However, they offer but glimpses of the popularity of certain kinds of animals with collectors and tantalizingly convey nothing about the value of the collection or its previous owners. Their frequency suggests that the desire for anonymity increased during the 19th century, to reach a climax in the present century. In the *Register* compiled by J. M. Chalmers-Hunt only seven anonymous sales contained zoological objects other than insects, shells, or birds and birds' eggs up to 1850, but from 1851-1900 there were fourteen, and from 1901-1920 there were fifty.

The great majority of 20th-century anonymous sales, and indeed of named collections, have been of mammals' heads and horns. This, and the frequency of sales of fish collections, shows an interesting response to the pressures of fashion and possibly economic events. Clearly both types of collection have little scientific value and are basically personal collections from hunting and fishing expeditions. Hunting and the preservation of trophies was probably most popular from the 1890s through to 1939. Both native British animals and the larger mammals of the British Empire, notably India and Africa, and to a lesser extent, Canada, were hunted with a view to taking the head, horns, or pelt. Not surprisingly, the salerooms through this period are dominated by sales of such trophies. Possibly economic factors influenced the sale of some collections, while in other cases sales certainly followed the death of the collector, for although many collections of natural history objects retain their interest after the death of the original collector, trophies are mainly esteemed by the hunter himself, serving to recall the highlights of the expedition. Few collections of this kind can be made today due to various circumstances, such as the very high cost of preparation of a head, and pressures for the conservation of many animal species which discourage or limit shooting.

Cased fishes as trophies have similarly been affected by economic and social pressures. Undoubtedly many early sales including fishes were of angling trophies, and they were of frequent occurrence. However, in the present century the cost of preparing fish trophies has risen astronomically and the majority of anglers today rarely kill the fish they catch. Possibly because of the smaller size of modern homes in which cased fish must be displayed such trophies were unpopular and virtually valueless in the immediately post-war period and specimen fish in a well-made glass case could be purchased for less than one pound at almost any dealer in such objects. However, with the change in taste since the late 1950s which has led to a greater esteem for the Victorian life-style, cased fishes have become popular and even hard to find. It is no surprise, therefore, that between 1969 and 1972 no fewer than four London sales have contained fishes. The pendulum of fashion has swung from a time when one could hardly give a cased fish away to a period when they command considerable prices.

BOTANICAL AUCTION SALES

WILLIAM T. STEARN*

The break-up and dispersal of a collection of objects purposefully assembled over many years, particularly a specialized collection, can rarely be regarded with other than mixed feelings if not with total regret, profitable though its sale may be to auctioneers and dealers. Such dispersal often brings to naught all the searching, expert knowledge and loving care which went to the making of the collection and robs scholars of the opportunity for comparison and study of related material thus gathered together. On the other hand, it may result in the enrichment of other collections being actively assembled for research or instruction or the pleasure of the public. This is generally true for books, pictures, ceramics and the like. Unfortunately, however, the break-up of a herbarium, i.e. a collection of dried plants scientifically named and annotated, is almost always a disaster for systematic botany. Problems created by the sale and dispersal in 1842 of the great Lambert herbarium continue to trouble taxonomists. It is horrific even to imagine how great would have been the catastrophe, how severely handicapped systematic botany would have been in Britain and indeed in many other countries, if the private herbaria of Sir Hans Sloane (1660-1753) and Sir Joseph Banks (1743-1820), now in the British Museum (Natural History), and of Sir William Jackson Hooker (1785-1865) and George Bentham (1800-1884), now at the Royal Botanic Gardens, Kew, had come to such an unfortunate end. Luckily most herbaria put on sale have been purchased as a whole and have sooner or later passed into the keeping of single institutions. The main period of the sale by auction of herbaria was between the first and last quarters of the 19th century. Thereafter this virtually ceased. The first half of the century saw the assembly of big private herbaria as research tools more or less adequate for the owners' needs. The second half saw their transfer by sale, gift or bequest to public institutions, as the growth of botanical exploration, the sheer bulk of the specimens collected, and the cost of housing them and paying the staff needed for their care and study made such institutions take over the functions of the earlier private herbaria. These had their origin in the 16th century.

Credit for inventing the practice of drying plants under pressure and fastening them on to sheets of paper for permanent reference purposes, thereby forming what was formerly called a *hortus siccus,* belongs to an Italian professor Luca Ghini (1490-1556). He taught at the University of Padua, which then drew students from all over Europe on account of its medical repute, and he evidently stimulated the more enterprising or the more acquisitive among them to make such collections of dried plants. The sheets of paper holding these specimens were bound into books and treated as such. Not until the 18th century did the advantage of having the sheets loose and stored in cabinets for study purposes become apparent; single sheets could then be re-arranged whenever views on

*Department of Botany, British Museum (Natural History), Cromwell Road, London SW7.

classification changed and they could, moreover, be compared by being placed side by side. This change from a static to a more dynamic and flexible system of herbarium-making was mainly due to the influence and example of Carl Linnaeus (1707-1778). His widely used encyclopaedic works provided a system of arrangement for specimens which could be continually revised and was relatively easy to understand. His students came to Uppsala from many countries and, on returning home, they either themselves set up herbaria or taught and encouraged others to do so. Public institutions owning such collections were few before the 19th century. Then these increased, as also did the number of wealthy and competent amateur botanists forming their own collections for study and the pride of possession. The curators of museums and the owners of private herbaria became buyers of specimens, competing against each other at auctions, and so created a market which earlier had hardly existed at all.

Financial support from institutions and individuals now enabled collectors to travel to regions little-known botanically and gather there sets of specimens on a large scale, often called *exsiccatae,* which were distributed to subscribers or offered for sale after provisional naming, usually with printed labels. They often included species new to science and greatly increased knowledge of plant distribution. By the purchase of such *exsiccatae,* by the gift of specimens in return for naming them, by the exchange of duplicates and sometimes by their own collecting in the wild, less often in gardens, leading botanists in the first half of the 19th century built up very large collections. The fate of these after the death of their owners depended on a variety of circumstances.

Most of the great private collections passed by bequest or gift into the custody of state museums or universities. Thus the herbarium of William Arnold Bromfield (1801-1851) became in 1853 the nucleus of the great Kew herbarium; that of Henry Barron Fielding (1805-1851) went to the University of Oxford, that of Charles Morgan Lemann (1806-1852) to Cambridge; the widow of Edward Rudge (1763-1846) gave his herbarium to the British Museum. Some others of great importance were, however, sold at auction and dispersed. Of their contents and extent the sale catalogues, some fortunately annotated at the sale with the names of the buyers of individual lots, provide almost the only available information. This may be valuable when the herbarium contained type specimens or other historic material because it may help in tracing their present location.

One of the earliest recorded sales offering botanical specimens was that of Margaret Cavendish Bentinck, Dowager Duchess of Portland (1715-1785); her herbarium was bought by Aylmer Bourke Lambert (1761-1842), whose herbarium was in turn sold by auction in 1842. Her chaplain the Rev. John Lightfoot (1735-1788) likewise had a herbarium which was purchased at the 1789 sale for Queen Charlotte and later dispersed, his plants coming ultimately to Kew and the British Museum. A number of such herbaria put up for sale have eventually reached these national institutions. Thus the herbaria of John Sims (1749-1831), sold in 1829, and of Robert Heward (1791-1877), sold in 1868, are both at Kew.

The two biggest herbaria split up by auction were those of Lambert and of Robert Graham. Lambert was a wealthy country gentleman with a keen interest in gardening and botany who, by a long series of purchases of *exsiccatae* and private herbaria, formed a herbarium estimated at his death to contain some 50,000 specimens from at least 130 collectors. Lambert delighted to have his collections studied by other botanists, especially if they could detect new species in them, as they often did. George Don the younger (1798-1856) based his monumental *General System of Gardening and Botany* (4 vols., 1831-1838) largely on material in Lambert's herbarium, of which his brother David Don (1799-1841) was the curator and librarian from 1820 to 1836. Lambert intended that this superb collection, so rich in type specimens, should go to the British Museum. Unfortunately his financial affairs had become greatly confused by the time of his death and he was then heavily in debt. The Museum, short of money, refused to buy it at the price requested, £2,500; it was accordingly put up for sale by auction in June 1842, lot by lot, 317 in all; it fetched as a whole £1,171 and Lambert's creditors received, out of the sale of the herbarium and his no less rich botanical library, fifteen shillings to the pound. The British Museum, with Robert Brown as buyer, purchased Buchanan Hamilton's Nepal herbarium (Lot 286) for £9, Pallas's primarily Russian herbarium (Lot 285) for £49 and the American herbarium of Ruiz and Pavon (Lot 103) for £270. Those parts bought directly by botanists such as Robert Brown, John Lindley, William J. Burchell, G. S. Gibson, Robert Heward, C. M. Lemann and John Miers can be directly traced, using the annotated sale catalogue at the British Museum (Natural History), and indeed are now divided among the British Museum (Natural History), Kew and Cambridge. Many small herbaria were, however, bought by the dealers Rich and Pamplin acting as agents for Continental botanists and institutions and cannot all of them be traced, despite the painstaking and remarkably successful investigation of the contents and fate of the Lambert herbarium by Honorine Miller, whose paper in *Taxon* **19**: 489-553 (1970) is a mine of useful information about the present whereabouts of Lambert material. The herbarium of Robert Graham (1786-1845), professor of botany at Edinburgh, was auctioned in 1846. It contained about 48,000 specimens and thus was almost as extensive as Lambert's though not so rich in type specimens.

Collectors such as Lambert and Fielding built up their herbaria primarily through acquisitiveness just as many of their contemporaries assembled collections of birds' eggs or butterflies without themselves making any notable contributions to knowledge by their study. Serious research-minded botanists such as William Jackson Hooker, Robert Graham, John Lindley, George Bentham and Charles Cardale Babington, however, bought and received specimens for their herbaria because they needed such material for their studies and because no public institutions then provided convenient and adequate facilities. Nowadays public institutions have grown so much, in their earlier stages by incorporating these great private herbaria, that no private individual can possibly compete with them or would wish to do so. The private collector's scope is more limited than that of his predecessors; he accumulates plants of a limited area or of a special group in which he is primarily interested. Such collections almost always pass by gift or bequest to institutions. The period when they had a commercial value making them suitable for sale by auction ended little by little as public institutions superseded the private herbaria for research purposes.

The sale catalogues of the mid Victorian period thus record an important stage of transition in the development of the most important tool of modern systematic botany, the institutional herbarium.

ORNITHOLOGY

CLIVE SIMSON*

In writing this introduction it is not my purpose to examine, in detail, any but a few of the Natural History sales, referring to ornithological matters, recorded in the *Register*. The sales are so tabulated that anyone interested can readily find out the whereabouts of a catalogue, or a source, referring to any particular sale. What I think is important is to outline the history of ornithological collecting and its results. To do this, one must examine the collecting of birds' skins, whether mounted or not; avian egg-shells; nests; as well as library material, in which must be included paintings, prints and photographs.

Birds were, of course, taken for the beauty of their plumage long before the date of the first sale recorded in the *Register*. The Red Indians of North America, with the feathered head-dresses of their chiefs, and the natives of New Guinea, with their adornments of the beautiful plumes of Birds of Paradise, family *Paradiseidae,* are obvious cases. There are many more. But this was hardly 'collecting' in the term we understand, though the procurement of certain birds' skins, such as the Huia, *Heteralocha acutirostris,* of New Zealand, by the ancient Maoris, comes very close. Huia skins were used as status symbols and for the ornamentation of the dead. They were often most carefully preserved as heirlooms and kept in beautifully carved caskets, known as *Papa-Huia*. The bird was never plentiful and that, sadly, it became extinct about 1907 was, assuredly, not the sole fault of the Maoris (cf. Buller's *Birds of New Zealand*.)

Birds could, of course, be readily taken by means other than shooting, as the pictures on the walls of ancient Egyptian tombs so clearly show. But it was the advent in the 18th century, of a reasonably accurate and serviceable gun, firing a light charge of pellets or dust-shot, that saw the start of serious skin collecting and with it, of course, much of our present knowledge. The first recorded auction sale of English and foreign birds' skins was that by Christie on 6 June, 1771, on behalf of Thomas Grace of London, when sixty-nine lots changed hands. By the middle of the 19th century the possession of stuffed birds became very popular. For instance, on 27 March, 1850, at a house sale by Harris & Belcher, 60 glass cases of rare British birds were sold by 'the executors of John Tomkins Esq of Abingdon'. Stuffed birds were usually tastefully posed in glass cases or, in the case of small birds of brilliant plumage, many would be displayed in one large, glass-sided cabinet. There was, probably, some minimal scientific value in these display cabinets but the data accompanying the specimens were often missing. It was in the work of serious skin collectors that the rapid advance of ornithological knowledge during the 19th century began. Such skins were seldom mounted but were preserved and lightly padded with cotton wool. They lie, curiously long, thin and unbirdlike, in row upon row in the drawers of Museums. In order to sort out juvenile and seasonal changes in plumage as well as sexual dif-

*Fyfield Grange, near Andover, Hampshire.

ferences and sub-specific differences, many skins of the same species were required. Looking at these long arrays from all over the world one wonders if a holocaust on quite such a scale was necessary.

However, there were many people keen to procure these skins. They were either wealthy men who liked the excitement of tracking down rare birds, either at home or in faraway places, or more frequently - men specially engaged and financed by a large museum. Such skins went straight into the museum's cabinets and have seldom appeared at auction. However, every so often there was a private museum sale, such as the formidable one of Bullock's Museum of Natural History in 1818, when a large number of skins was auctioned. In such cases public museums were frequently among the bidders. It is worth giving some details about the Bullock sale, which took place on twenty-three separate days, between 29 April and 11 June, 1819, in London. William Bullock, a Liverpool jeweller and goldsmith, whose private collection of 'natural and foreign curiosities' was housed partly in Liverpool and partly in London, was himself listed as the auctioneer. Some remarkable objects were included for it was at this sale that a skin of a previously unrecorded petrel, which had been shot by Bullock off St. Kilda in the summer of 1818, was bought by the British Museum. Dr. Leach, an assistant zoologist at the British Museum, identified it as *Oceanodroma leucorrhoa* and the bird is known, to this day, as Leach's Petrel. It is not inappropriate that such a name should have stuck. Mr. Bullock had missed immortality by a pin-feather! The British Museum also bought the skin of a Great Auk, *Alca impennis,* sent to Bullock in 1813 from a bird killed in the Orkneys that year. There is only one other such skin from British waters and that is in the Trinity College Museum, Dublin, taken in 1834. At the same sale another most interesting event took place. I am indebted to Sir Hugh Elliott of Oxford for this information and quote his letter of 19 November, 1973, verbatim: '(It was at this sale) that three specimens of the 'Island Hen', or moorhen of Tristan da Cunha, *Gallinula nesiotis* changed hands. The specimens subsequently vanished from sight, although they could well exist in a private collection or small provincial museum. The interest of this lies in the fact that the species has been considered extinct for upward of a century, but *may* have been rediscovered this year (1973). At the moment there are only two fully authenticated specimens in the world's museums, so it would be of the greatest scientific interest .... if even one of these skins could be found.' (A further reference to this matter is given in *Bulletin of the British Ornithologists' Club,***92**, nos. 3 & 4.

No consideration of the collecting of birds' skins would be complete without the name of William Hancock. This gentleman amassed, in the mid-19th century, at first privately, and later for the museum in Newcastle, Co. Durham, to which he gave his name, a collection of mounted skins of the Raptores, unequalled in Europe. The display is still there for all to see. Great birds, such as the White-tailed Eagle, *Haliaetus albicilla*, Golden Eagle, *Aquila chrysaetos*, Osprey *Pandion haliaetus* and Peregrine, *Falco peregrinus* and many more, are superbly mounted, in suitable landscapes. I cannot find that Hancock ever bought second-hand at auction. He depended on two types of people for his material; the wealthy, amateur naturalist, such as Charles St. John, and hard-headed professionals, such as the brothers William and Lewis Dunbar. St. John provided him, free, with specimens of Golden Eagle, Sea-Eagle, Osprey and Peregrine. St. John has been unjustly accused of causing the extinction of the Osprey in Scotland, nevertheless, with the able assistance of the Dunbars, he certainly caused the extinction of this fine bird in Sutherland. All his trophies, skins and eggs went to Hancock. For one year Wolley, before he departed to northern Scandinavia and fame, also harried the larger raptores nesting in northern Scotland. Hancock, himself, in 1850, joined Lewis Dunbar on Speyside and insisted that Dunbar should shoot the Loch Morlich hen osprey as she circled the eyrie. Even hard-bitten Dunbar regretted this act.

One must not give the impression that bird-skins are collected only by museums. There are many collections still in private hands and their owners were regular attenders at the auctions at Steven's Rooms, until this firm ceased to operate in 1939. In the period from 1905 to 1914, there used to be quite a clique of quasi-sporting, quasi-scientific, bird-shooters, who parcelled out those bits of the East coast of England considered most likely to produce rare birds at migration times. The late Lt.-Colonel J. K. Stanford was one such, and, in 1912, on the East Coast, he shot and skinned two Barred Warblers, *Sylvia nisoria;* one Red-breasted Flycatcher, *Muscicapa parva;* one Firecrest, *Regulus ignicapillus;* one Blyth's Reed Warbler, *Acrocephalus dumetorum* (the first ever recorded in Britain), and two Bearded Tits, *Panurus biarmicus.* I quote this, because of Colonel Stanford's, and other skinners', inveterate hostility towards oology and oologists.

The sale of the mounted Great Auk, *Alca impennis,* at Sotheby's for £9000, on 4 March, 1971, appears to be the highest price ever paid at auction for a bird's skin in Britain. It was put up for sale by Baron Raben-Levetzau of Denmark, and was bought by Dr. Finnur Gudmundsson, Director of the Natturugripasafnid at Reykjavik, acting on behalf of the Icelandic Government. The bird was taken (*c* 1821) in Iceland.

On 17 March, 1972, at Sotheby's, some skins of the Passenger Pigeon, *Ectopistus migratorius* were offered for sale by the Bankfield Museum, Halifax, Yorkshire. I am indebted to Richard H. Pough, in his *Audubon Bird Guide,* for the facts of the amazing story of this extinct North American species. Probably no other bird has ever occurred in quite such huge concentrations as the Passenger Pigeon; their vast migratory flights darkened the sky and took hours to pass. One great nesting in Wisconsin, in 1871, covered 850 square miles and contained 136 million nests. As late as 1878, one and a half million birds are known to have been sent to market by railroad alone, from the last great nesting. They made very good eating. The last wild specimen was shot twenty-one years later, in 1899 and the last survivor, a bird hatched and reared in captivity, died in Cincinnati Zoo in 1914, aged twenty-nine. Thus, from flights estimated to consist of about two billion birds in 1860, no living member of this species existed after 1914. A frightening story indeed. The moral, if any, is that no species, man included, can rely solely on numbers for its continued existence.

In reviewing the particulars of auctions in the *Register* one is immediately struck by the great frequency of oological material offered for sale. There can be no doubt that many people, over the long period of time covered in this work, have been desirous of possessing an egg collection; or of adding to an existing collection by purchase. One must say that, if specimens are properly treated in the first place, then avian shells are as indestructible as sea-shells, provided they are kept free from mould and mites and not exposed too frequently to strong light. If these conditions obtain, then they will retain their beauty of colour, marking and form indefinitely. Just as the early conchologists collected sea-shells, primarily for their beauty or rarity, so did the earlier oologists. In both cases, the mid-19th century saw a surge in the desire to acquire such objects of natural beauty. This desire to collect is a remarkable phenomenon in the human race. At the higher price levels, it becomes a hedge against inflation. In an age when many people are dying of starvation, a piece of oil-daubed canvas can fetch a million pounds; an egg-shell can be worth £1,000. If only the shell was full of yoke and albumen, then one human life might be extended by a few days! But this is not the way things are - one must review facts. The facts are that people, given money and leisure, turn to some absorbing hobby, ranging from collecting old-masters to match-boxes. From the mid 19th century many people in Britain possessed both wealth and leisure: a few of them became oologists and, in the process, considerably advanced ornithology.

Early egg collections usually consisted of single eggs, blown with a hole at either end, and frequently stuck with glue to a sheet of stiff paper, before being placed in a cabinet drawer. Some country folk strung eggs together through the blow-holes, and hung these brightly coloured chains in their windows. There can be no doubt that the beauty of many avian shells played a large part in the desire to collect them. Later on, collections became far more sophisticated, with whole clutches being collected and placed in compartments in drawered cabinets. They were blown with only one hole in the side and full data records were kept. Eggs varied so much, in both colour, size and shape within the same clutch that it was considered of scientific importance to show the full laying; added to which many nests were perforce damaged in taking even one egg, leading to the desertion of the nest. It was furthermore argued that most birds replace a lost clutch very quickly, and the taking of one or two eggs would, in fact, tend to reduce the level of young eventually fledged.

It is probably true to say that egg-collecting has not, in itself, caused damage to any species, unless that species was in jeopardy through other causes. When a bird became rare it occasionally happened that professional egg dealers (and there were a few) stepped in. Not only was the first clutch taken, but the repeats were taken also, with the knowledge that the rarity value of each succeeding set was that much greater. This is similar to the present view that just a few persistent dealers were responsible for the extinction of the New Forest Burnet Moth, *Zygaena meliloti,* in 1926. However, in fairness to scientific oology, I think it is now generally agreed that the gun has always proved a far greater threat to birds than egg-collecting, viewed on a world-wide scale, and that the destruction of natural habitat far outweighs either.

The sales of some large egg-collections, such as that of Percy Bunyard, at Steven's Rooms on 7 February and 7 March, 1939, are recorded in the *Register;* but the most important collections never came under the auctioneer's hammer. I refer to the world-wide collection of the late Hon. L. W. Rothschild (later Lord Rothschild); to the palaearctic-species collection of the late Rev F. C. R. Jourdain; and to the collection of purely British species of the late Edgar Chance. The Rothschild collection became part of the private Tring Museum (now taken over by the British Museum (N.H.). The Jourdain collection was picked over by some thief during the Second World War and many of the most important specimens were lost. The residue was acquired by the dealer Gowland. The Chance collection was bought by the British Museum for a sum in the region of £8,000, shortly after the Second World War, which took the wind out of the sails of those who said that egg-collections contained little of scientific value.

One article, the egg of the Great Auk, *Alca impennis,* figures very frequently in the various auctions listed in the *Register*. The appearance of the eggs of this species always excited interest. The Great Auk was a large, flightless member of the Razorbill genus, which stood, penguin-like, a full thirty inches tall. It laid a single egg measuring on average, 124 by 77 mm. The egg is strikingly beautiful, being pyriform, with a ground colour ranging from bluish to creamy-white, blotched, spotted and streaked with varying shades of dark brown to black. In the 18th century it was conservatively estimated that one and a half million birds nested on the islands of Hamilton Inlet, Labrador, alone. There were also large nesting colonies off Newfoundland and Iceland. The wintering flocks, off Florida, were once counted in millions. Yet by 3 June, 1844, there was only one pair left, plus an embryo within its egg-shell. Allan Eckert in his book *The Last Great Auk* says 'The two stately birds waddled ashore on Eldey Island off the south coast of Iceland. Soon afterwards the female laid an egg there. Fate has always been fickle, but it was seldom more capricious than on 3 June, 1844.' (A boat, carrying six men and three boys, landed on Eldey from a ship anchored off-shore. The men started clubbing the crowds of birds, mostly murres, too slow to leave their nests.) 'Suddenly one of the men glanced up and spied the huge forms of the two Great Auks. "Garefowl" he screamed. Swiftly the men spread out and advanced on the two birds with their clubs at the ready. The birds, on land, were sluggish and awkward. The clubs streaked down. One of the men, intent only upon the Great Auk's fleeing form, stepped on the single, large egg and crushed it. He had just destroyed the last of a species. Thus it was that on 3 June, 1844, the Great Auk became extinct from the face of the earth.'

There are, at this time, about eighty skins and seventy-five eggs of of the Great Auk in existence. The *Register* frequently mentions the occurrence of a Great Auk's egg at auction. The value of a Great Auk's egg depends on the markings (some are almost unmarked — just a dirty white) and on its condition. The seamen, who originally took these large eggs, often knocked the small end off and used the egg as a drinking cup! In 1966 the Trustees of the British Museum (N.H.) published a work entitled *Eggs of the Great Auk* by P. M. L. and J. W. Tomkinson. This book includes a photograph of every known existing egg and a summary of its provenance and changes of ownership up to 1965. Each egg has a number. It also contains most interesting details, largely culled from the catalogues listed in the *Register*. For instance, at the Bullock sale of 1819, already referred to, a Great Auk's egg was bought by the British Museum for £16.15.6d, while at the same sale it also acquired another Great Auk's egg for 12s. or 17s. By 1880 the price had risen sharply, Lord Lilford paying £107.2s. for an egg at the Stevens' auction sale on 2 July of that year. On 17 April, 1902, an egg changed hands at Stevens' Rooms for £252. On 14 November, 1934, Captain Vivian Hewitt bought an egg at a Stevens' auction for £315 and two other eggs were bought at the sale by Rev. F. C. R. Jourdain for £147 and £220.10s., which were not such good specimens as that bought by Hewitt, who, nevertheless, eventually acquired the Jourdain specimens at the dispersal of the Jourdain Collection. In December 1939, Gerald Tomkinson, the father of one of the co-authors of *Eggs of the Great Auk,* paid £400 for an egg at a private sale. This egg, a fine specimen, had previously changed hands at Stevens' Rooms, on 29 October, 1901, for £252. Finally, it is worth recording that Hewitt bought three eggs, through the dealer Rosenberg, for £1,000 on 19 June, 1946.

The name of Vivian Hewitt looms large in any consideration of the eggs of the Great Auk. At the time of his death. on 18 July, 1965, he possessed twelve eggs of this species, while the British Museum had only six. The only real competitor was the Museum of Comparative Zoology, Harvard College, U.S.A., which owned ten. Between them, they owned nearly forty per cent of the existing eggs of the Great Auk. As soon as Hewitt started bidding for one of these eggs his competitors knew that he would go the limit. He was determined to possess it and, in the process, may well have pushed the price beyond reason.

What is one to make of this compulsive hoarder of Great Auk's eggs? Hewitt was a man of great wealth and tremendous vitality, yet, after an adventurous youth, he spent the last thirty years of his life largely as a recluse in Anglesey, North Wales, or in the Bahamas. During this time he amassed, chiefly by purchase, the largest collection of birds' eggs ever privately owned in this country and, probably, anywhere else. And how he collected! As well as birds' eggs and bird-skins, he collected butterflies, stamps, coins, model steam-engines and vintage cars (an immaculate 8-litre Bentley gleamed in the middle of his entrance hall.) Besides the twelve eggs of the Great Auk, he possessed two far finer mounted Great Auks than the specimen that fetched £9,000 at Sotheby's in 1971. His egg-collection room was stacked with cabinets and boxes. many still in the unopened packing-cases in which they had arrived. As many specimens were subsequently found to be without data, it is very difficult to assess his stature as an ornithologist.

Hewitt's egg-collection, less the Great Auk's eggs, was presented to the British Trust for Ornithology, and four large pantechnicons were needed to carry it from Anglesey to Tring. A rough estimate by Dr. J. J. M. Flegg, the Director of the Trust, put the number of eggs at half a million, and the skins in the region of 'at least tens of thousands.' These eggs and skins had a world-wide provenance. (cf: *The Modest Millionaire* by W. Hywel.)

The Great Auk's eggs, which Hewitt had kept in the Bahamas, were sold by his Trustees to American interests with the exception of egg No. 56 (Tomkinson number), which was acquired by the National Museum of Wales. A considerable revision of the Tomkinson list is thus necessary. It would be interesting to know the prices paid.

I may have given the impression that the Great Auk's egg is the ultimate in avian shell value. Such is not the case. The *Register* records that at the same 1971 Sotheby sale as saw £9,000 given for a Great Auk skin, £1,000 was given for an egg of the extinct *Aepyornis maximus*. This bird, which was flightless, inhabited Madagascar and was the largest known avian species to have trodden the earth. It stood nearly ten feet high, weighed half a ton, and laid an egg a foot long. This egg held more than two gallons, and thus had eight times the volume of an ostrich egg. Monsieur Isadore Saint-Helaire, of the French Academy of Sciences, named it *Aepyornis maximus* in about 1850. It is unlikely that any European ever saw this bird alive, and, though the date of its extinction is not known for certain, it probably followed a few hundred years after the arrival of the first human beings on the island, which is estimated to have been between 2,000 and 1,500 years ago. Undamaged eggs were, and still are, found after the heavy seasonal rains have washed them out of the soil in certain parts of the island. Of interest is that a recent X-ray has shown the bones of a well-developed embryo within an egg (cf. *National Geographic Magazine,* October, 1967).

In summing up the activities of skin-collectors and egg-collectors, it is appropriate to say that to the first we owe the proper identification of avian species, their seasonal changes of plumage, their juvenile appearance, their sub-specific differences and average measurements obtained from many skins: to the second we owe our knowledge of bird behaviour at its most intense expression. This is because the nest is the focus of the whole mating ritual. Plumage is at its brightest, displays are frequent and, above all, the voices of birds are at their most exuberant. It is not necessary to collect eggs to record these facts. It happened, nevertheless, that it was through egg-collectors, such as the Rev. F. C. R. Jourdain and J. Walpole-Bond, that much of our modern knowledge was obtained. In order to find nests they had to learn a great deal about birds. They recorded all that they saw and heard, and so ornithology progressed. Furthermore, the clues to be obtained from eggs must not be forgotten: the scientific study of eggs, - oology - has much to offer the whole ornithological inquiry. British collectors have travelled the whole world in pursuit of their hobby; Seebohm to the Petchora and Yenesei valleys in Siberia, Wolley to Lapland, Rothschild to South America, Stuart-Baker to India, particularly the Himalayas, Congreve to Spitzbergen and Iceland and Colebrook-Robjent to Uganda and Zambia. There is nothing parochial in their findings. All discovered something new and all were collectors.

The collecting of birds' nests provides a great deal of information but unfortunately they are not easy to preserve and there are problems of space as well. When nests did appear at auction, it was not only museums that bought them. Certain nests sold readily, such as those of the Weaver Birds of Africa, since they were fairly easily preserved and had some decorative value.

Finally, there is library material. The *Register* shows that as long ago as May 1808, there was interest in pictures of birds for, at a Sotheby sale, of that date, 1,800 drawings and prints of birds changed hands for £378. There can be no doubt that more books, many with superb illustrations, have been published on ornithology than on any other Natural History subject. This is because birds have an almost universal appeal and because their beauty of flight and plumage lend themselves readily to the artist's attention. It is not surprising that some of these books should nowadays fetch quite staggering sums at auction. For instance, David Evans of Fordingbridge, in Hampshire a specialist in bird books, paid £10,000 for a Daniel Giraud Elliott monograph of the phasianidae. He also paid £10,000 for John Gould's *The Birds of Great Britain* at the same sale, which took place, under the auspices of Christie's, at Croxteth Hall, Liverpool on 20 September, 1973. A few days later I noted that a Gould had changed hands for £8,000 at an auction by Sotheby's in London but the buyer was not mentioned. Shortly afterwards my friend, David Evans, visited me on a matter connected with my small library, and I could not resist gently pulling his leg by mentioning that he could have got a Gould for £2,000 less than he had paid in Liverpool. He smiled and said that it was he who had bought the £8,000 copy too. He further told me that bird books of this quality and rarity had increased at least tenfold in value during the last ten years.

I have before me, as I write, a first edition of Bewick's two volume *History of British Birds,* published 1797 and 1804, priced at 15s. a volume. Each contains charming wood engravings, not only of birds, but also of country scenes. On page 285 of Vol. I is a neat little vignette of a garden privy. In the first impression of the first edition an old lady is depicted seated in this rustic convenience. This is known as the 'naughty edition' because in the second impression the old lady was blotted out with ordinary ink after printing. It was clumsily done and cut a quarter off the value of the book, at current prices. Alas, no old lady is visible through the ink cloud in my copy. Bewick's pictures were wood engravings and were not coloured. However, many of the early bird books contain hand-coloured pictures of birds. Though in some cases the letter-press may be undistinguished, the pictures are much sought after nowadays. All in all, these old bird books give a great deal of pleasure and, occasionally, instruction. They are well worth a collector's interest, as the catalogues listed in the *Register* show.

It is early days yet to decide on the relative artistic merits of the many fine bird photographers, whose contribution to ornithology cannot be ignored. In Britain, I suppose it is fair to say that the work of Eric Hosking and G. K. Yeates has set a standard. Their beautiful pictures grace so many books about birds that they cannot be left out of the collector's consideration. Abroad, the same considerations apply. Whether any of them will reach the stature of a Gould or an Audubon remains to be seen. Nevertheless, photography is an art; and birds have graciously lent themselves to yet another visual interpretation of Man's absorbed interest in Nature.

One cannot conclude without reference to the Protection of Birds Act, 1954. Under the terms of this Act it is illegal without a licence to sell, offer for sale or have in one's possession for sale an egg (including a blown egg) of a wild bird of any species, if any bird of that species has nested in the British Isles in a wild state. (One must note that the Great Auk once bred in the British Isles and its eggs are therefore subject to this Act.) It is also illegal to sell the skin or plumage of any wild bird, except certain 'sporting' birds, specially mentioned in Schedule 3 of the Act, except by licence or proof that the skin or plumage was lawfully imported. It is as well that all concerned are aware of the terms of this Act. The *Register* records that on 6 November, 1969 Messrs. Hall and Palmer, at a house sale, were summonsed by the R.S.P.B. for selling thirteen eggs, all of which were at least fifty years old. 'It seems doubtful if a case like this has ever been heard before in this country,' the magistrates declared. The defendants were granted a conditional discharge.

Thus, from 1954, the sales of skins and eggs of wild birds recorded in the *Register,* from 1771 onwards, show a marked decline. The sale of the eggs of most palaearctic avian species is finished, unless with a licence. The sale of *all* bird-skins is likewise ended. In the past, much of real scientific, ornithological value changed hands at the auctions listed in the *Register*. Here, however, is recalled the last of private enterprise. In an overpopulated world one must suppose this was inevitable, but let us not forget that it was the individual collector, however misguided he may be thought to have been today, who produced our present knowledge which, incidentally, is by no means ultimate. There is much yet to be learnt.

REFERENCES

Bewick, T. 1797. *History of British Birds*. Newcastle
Buller, W. L. 1888. *A History of the Birds of New Zealand* (2nd edition). London
Bulletin of the British Ornithologists' Club, **92**: Nos. 3 & 4
Campbell, B. & Ferguson-Lees, J. 1972. *A Field Guide to Birds' Nests*. Constable
Check List of New Zealand Birds, 1953. Wellington, N.Z.
Dresser, H. E. 1910. *Eggs of the Birds of Europe*. London
Eckert, A. 1964. *The Last Great Auk*. Collins
Entomologists' Gazette, **17**: 187 et seq
Harvie Brown, J. A. & Buckley, T. E. 1895. *Fauna of the Moray Basin*.
Hywel, W. 1973. *The Modest Millionaire,* Gwasg Gee
National Geographic Magazine, October 1967

Newton, A. 1864. *Ootheca Wolleyana,* London
Pough, R. H. 1951. *Audubon Bird Guide,* Doubleday, U.S.A.
Protection of Birds Act, 1954. 2 & 3 Eliz. 2 ch. 30
Seebohm, H. 1901. *The Birds of Siberia,* John Murray
St. John, C. 1850. *Sportsman & Naturalists' Tour in Sutherland,* London
Tomkinson, P. M. L. & J. W. 1966. *Eggs of the Great Auk,* London
Witherby, H. F., Jourdain, F. C. R., Ticehurst, N. F. & Tucker, B. W. 1938-41.
 The Handbook of British Birds, Witherby

FOSSILS

W. D. IAN ROLFE*

Although commonly sold interspersed with shells, corals and minerals, as in the Portland and Lever sales, fossils only occasionally fetched prices comparable with those commanded by shells. They seem to have been less popular objects for the collector, despite their often greater rarity, largely on account of their inherently less aesthetic appeal, a fact attested by the comparative rarity of fossils in still-life paintings. Auction sales have thus not played such a significant part in the development of palaeontology, and there is no palaeontological equivalent here of the taxonomically valuable Portland catalogue.

As early as the 1660's (but not published until 1705), Robert Hooke advocating the collection of fossils for the Royal Society's Repository stated that 'the use of such a collection is not for divertisement, and wonder, and gazing, as 'tis for the most part thought and esteemed, and like pictures for children to admire and be pleased with, but for the most serious and diligent study of the most able and proficient in natural philosophy'. These sentiments were frequently echoed later, notably by Woodward (1729) and Owen (1862), who wrote: 'A museum of nature does not aim, like one of Art, merely to charm the eye and gratify the sense of beauty and of grace... its .... purpose is to impart and diffuse that knowledge which begets the right spirit in which all Nature should be viewed, there ought to be no partiality for any particular class, merely on account of the quality which catches and pleases the passing gaze. Such a Museum should subserve the instruction of a People; and should also afford objects of study and comparison to professed Naturalists, so as to serve as an instrument in the progress of Science'.

William Smith's 1795 discovery that (often fragmentary) fossils could be used to identify stratigraphical succession helped to reduce undue emphasis on the purely aesthetic fossil. As Konig reported to the British Museum Trustees in 1817, when recommending the purchase of Smith's second collection of fossils (260 specimens of 120 species for £100): 'Their value, considered as specimens is very small; but it appears to be otherwise with regard to the value derived from their utility when combined with the strata in which they have been found, and the advantage which may accrue from them to the proprietors of land, by their pointing out the nature of the strata and disclosing in consequence the mineral riches which in multitudes of instances the Proprietors of estates are in no degree apprized of. The value of a collection of this kind, as it cannot be calculated according to the selling price of the articles of which it consists, must be derived from a consideration of the labor and ingenuity of the collector and from the utility of the collection to the landholders of Great Britain' (Eyles 1967). Smith was only temporarily rescued from the debtors' prison by this last sale: he was to spend ten weeks there in 1819. With many demands on their funds,

*Hunterian Museum, The University, Glasgow.

museum curators often drove hard bargains, taking advantage of their would-be seller's predisposition to sell to a renowned institute.

'varietas delectat' was the appreciative remark made of the encyclopaedic character of a typical 17th-century collection. 'Such collections were regarded as a means of extending knowledge as well as offering enjoyment to leisured people who wished to while away their time. Collecting was recommended as a medicine against the sin of laziness, 'idleness creates wickedness'. Many early scientists were naturally among these collectors, but the scope of their interests transcended their professional skill. They collected archaeological and ethnological specimens as well as specimens of nature, and represented to some degree the type of the 'complete man' in the humanist sense of the word (Wittlin 1949). As S. P. Dance (1966) has said, 'during the late 17th and early 18th centuries a cabinet of natural and artificial curiosities was considered as indispensable to the well-appointed household, mansion or palace as was a collection of art treasures; and large sums of money were squandered on both indiscriminately. Often, the cabinet contained specimens which would now be considered worthless, although the owner may have given more for them than he gave for works of art which would be priceless today. Some collections were formed by men genuinely interested in natural history, some by men who considered that a cabinet reflected their social status, financial position and intellectual attainments'. This was the age of the dilettanti; the miscellaneous character of such collections is illustrated by the following lots from the painter Edward Steele's 1759 sale:

6, Sundry fossils &c in partitioned drawers;

13, An eagle stone, an ostrich's egg, a carnelian, a millepedes in spirits, a pair of wooden shoes, an indian firework.

15, A Laplander's shirt... a hair cap of Mary Queen of Scots.

One of the few oil portraits featuring fossils, towards the end of this period, is the rather dull portrait (now in the National Museum, Stockholm) of Thomas Haviland of Bath, by Gainsborough c. 1761. It shows a typical 18th-century physician with the latest edition of the current medical text in his hand, and with an arietitin ammonite and *Pseudopecten* on his bookshelf. The finest collectors of this era of disciplined curiosity are, of course, exemplified by Tradescant, Sloane, Mead and Balfour:

View there an urn which Roman ashes bore,

And habits once that foreign nations wore,

Birds and wild beasts from Afric's burning sand,

And curious fossils rang'd in order stand' (in Murray, 1905, p.176).

The subsequent growth of the spirit of inquiry is well illustrated by the history of palaeontology. The controversy over extraneous fossils (so-called to distinguish them from 'native' fossils, i.e. minerals and rocks), whether they were natural, organic remains or the result of inorganic growth *in situ* or other mechanisms, was a central question that engaged the best minds of the day, including Voltaire, and led to the amassing of collections by major seats of learning that they might contribute to the debate. (See Jahn 1963, p.163-8 for a summary of cabinets.) Lhwyd's catalogue of the Ashmolean Museum fossils (*Lithophylacii Britannici* 1699), besides being the first catalogue of British fossils and an important illustrated work much used by his contemporaries, contains his correspondence with John Ray on this subject. The issue of questionnaires and instructions to collectors, for example by Tradescant 1625, Boyle 1666, Plot 1670's, Petiver 1690's and Woodward 1696 epitomised the age of scientific revolution, although this successful device was used as late as 1853 by Agassiz. Fossils played a crucial role in the recognition of the immense age of the Earth and in the development of evolutionary modes of thought. The emphasis was on the significance of fossils, not their beauty. Woodward, referring to a specimen of dendrites in 1725 wrote: 'Tis pity a gentleman so very curious after things that were elegant and beautiful should not have been curious as to their origin, their uses and their natural history; about which he was little solicitous'.

Serious collectors preferred to give, exchange or sell their specimens to contemporary scholars or museums. Thus Woodward (1725) prized his 'piece of the true Ludus Helmontii found in the place mentioned by that author (the first topotype?) near Antwerp, and brought thence by his son Fr. Merc van Helmont into England. He gave it to ....and he to Mr. Kemp: of him I had this piece'. Another specimen arrived via Sir Isaac Newton. This concern for adequate distribution of comparative materials could be taken to lengths that would alarm curators today: 'two pieces of a large tooth, being broke off from one kept in the Duke of Wirtemberg's Musaeum, and presented by the keeper to Captain Richard King, who gave it me' (Woodward 1725). Woodward in his turn sent a fossil cidarid spine to Balfour (1697); he also bought the whole of Scilla's important collection, and many other collectors, including Sir Hans Sloane and William Hunter, similarly acquired whole collections. In 1665-6, shortly after its foundation, the Royal Society purchased Robert Hunt's collection containing fossils for £100.

This tradition of offering collections to the few museums then in existence was established early on. John Woodward, author of numerous works (see Eyles 1971), including *Natural history of the fossils of England* (1729), *Fossils of all kinds digested into a method* (1728), and *Natural history of the Earth* (1695), bequeathed his important collection of English fossils to the University of Cambridge in 1727, where they are preserved today with their original cabinets. Only his 'additional' English fossils and his foreign fossils were scheduled for auction with his books and other objects in 1728. They are not listed in the sale catalogue however, and it seems likely that they were not auctioned, but that a private sale was negotiated, the University of Cambridge sanctioning payment of up to £1000 but acquiring them for £500 (cf. Eyles 1971).

The auction was viewed as a last resort by the major collectors, the classic example being Sir Ashton Lever, whose collection was auctioned in 1806 and the specimens, including many fine fossils, were spread over several collections (cf. Smith, 1960). Fortunately, annotated copies of the catalogues (particularly Laskey's at Middleton Library, Manchester) enable particular specimens to be traced. A manuscript catalogue of the Leverian fossils was made by Emanuel da Costa, 'for which Sir Ashton paid the Author 200 guineas'. This was auctioned as lot 43 in the anonymous sale of 1832.

Only rarely have museums been driven to such ends. Lhwyd actually sold off duplicate specimens, accompanied by his 1699 *Lithophylacii,* to subsidise his travels and later generally to augment the scanty resources of the Ashmolean. It is true though that the lack of funds, or the niggardly way in which they were dispensed, occasionally drove collectors away from museums or forced them to make alternative arrangements. Thus Sir Hans Sloane thought that none of the existing institutions was fit to receive his collection (containing 1275 fossils). He would only permit it to be offered at £20,000 in turn to The Royal Society, Oxford University, the College of Physicians at Edinburgh and then to four European museums, and finally put to auction if Parliament would not pass an

Act to purchase the collection at this same price. Which of course it did, to found the British Museum.

Evelyn records that by the late 17th century dealing in curiosities had become a recognised calling. Coffee-houses were filled with objects of curiosity. The 'natural rarities' from the 'Musick house at the Mitre, near St. Paul's' in 1664 eventually came to the British Museum. Sloane gave specimens to Don Saltero's Coffee House museum in Chelsea (*fl.* 1710). The old curiosity shop in Camberwell survived to the 1860's as a relic of this earlier trade, famous fossil shops being Mary Anning's in Lyme Regis during the early part of the 19th century, and Mawe's (followed by Tennant's) in the Strand at the same period. Prior to the establishment of the learned societies, the auction rooms continued this social function by affording 'to collectors and men of science the opportunity of meeting, of knowing one another and conversing'.

But, as Lord Rothschild points out (Allingham, 1924), curators could not always be at sales, and 'often unique specimens have been dispersed for a mere trifle, and fallen into the hands of ignorant curio hunters at whose death, or transmigration elsewhere, they are eventually destroyed and lost forever to science'. In 1830 for example a Great Auk fetched 8s. (perhaps £3 today) whereas 'a hawk' fetched 6s: an interesting contrast to the figure of £9,000 paid for the Great Auk sold at Sotheby's in 1971. (Some idea of the relative values of money is obtainable from Cleevely 1974 p. 473). Mantell was able to buy good fossils 'all very cheap' at Thomas Hawkins's (of *The book of the Great Sea-Dragons* fame) sale in 1844, where 'not more than 25 persons were present, and none of the valuable specimens were bid for' (Curwen 1940). That curators did attend, or delegated their authority to 'buyers in', is clear from the annotations in some of the catalogues. Thus we find Clift buying for the Hunterian Museum of The Royal College of Surgeons, London at the 1806 Leverian sale. Clift is also noted, with Konig of the British Museum, as buying many of the fossils at the 1819 sale of Bullock's London Museum of Natural History ('unquestionably the most extensive and valuable in Europe' according to the catalogue), in competition with dealers such as Mawe. Mantell's journal records Konig paying 145 guineas (£147 in the British Museum's copy of the catalogue) for a young mastodon cranium from Big Bone Lick, at the sale of Finnell's collection in 1836. Clift's copy of the anonymous 1832 sale is preserved, and he was also at the sale of the collection of Professor Webster (of University College London) in 1846, this time in competition with Mawe's successor, Tennant. His copy of the catalogue notes that 'a cart-load of barrels, boxes and baskets full of geological specimens' was bought for 19s. 'for Raile-Roads, as a Witty Purchaser, a Mac-Adamiser, chose to express himself'. Webster's collection was sold on behalf of the Crown, but fetched only trifling sums, so that Clift wrote 'the labours of poor Webster's anxious life will hardly produce a pair of embroidered military pantaloons for Prince Albert'.

The use of such annotated catalogues enables specimens to be tracked down from sale to sale to their latest buyer, who hopefully is to be found in the indices of final repositories by Woodward, Sherborn or Cleevely (MS). The echinoid '*Coronula diadema*', figured by Parkinson in his 1811 *Organic remains of a Former World*, was bought at the sale of the latter's collection in 1827 by T. Image, repurchased at Image's sale in 1856 by W. N. Last from whom it was obtained by the British Museum. Occasionally sale catalogue entries may be correlated with penecontemporaneous inventories of the museums that purchased them. Many of Laskey's entries in his own 1813 account of the Hunterian

Museum are lifted verbatim from the 1806 Leverian catalogue. However, as Murray discovered, such early museum accounts may be even rarer than auction sale catalogues: few copies of the *Companion to the Leverian Museum,* 1790, or to Bullock's Museum, *c*1810, are known. Very occasionally the catalogues contain locality or other details that may be unobtainable elsewhere. The Birch 1820 catalogue, for instance, has a frontispiece of 'a crustaceous insect' which must be the earliest representation of the Liassic eryonid decapod *Coleia,* hitherto unrecorded by palaeocarcinologists. The 1842 McEnery catalogue notes that 'all specimens marked "drawn" are specimens from which plates of Buckland and Cuvier have been taken' (largely Kent's Cavern Quaternary mammals).

Clearly, many important fossils have changed hands at auction sales. Thus Sowerby purchased at several sales. He is noted as buying Lot 93 'shells, petrified wood, Gum &c found in making the archway at Highgate' at Bullock's 1819 sale (doubtless to add to the collection of specimens from this locality that he had earlier been given by Bullock himself). Sowerby also bought several of James Parkinson's fossils at the 1827 sale (no copy of this catalogue is yet known extant). Some of these later figured in Sowerby's important work *The mineral conchology of Great Britain* (1812-46), the first work to provide adequate illustrations of localised British fossils, and the basic source of nomenclature for many fossils. Many more were obtained by donation or loan, however, and by naming fossils after their donors the Sowerbys immortalised them by a reward system that continues today. Donors have long cherished such recognition, as Hearne parodied Sloane,

It is my wish, it is my glory,
To furnish your knick Knackatory.
I only beg that when you show 'em,
You'll tell the friend to whom you owe 'em.

The Sowerby collection was offered for private sale in 1831, but remained unsold until bought privately by the British Museum in 1861 for £400. This was the lowest figure that Sowerby would accept, he having valued it at £450, a difference which seems to have persuaded J. de C. Sowerby not to part with the whole collection (Cleevely 1974).

The pattern of sales of fossils represented in this register reflects that of natural history as a whole - most activity taking place at the beginning, middle and end of the 19th century, with a minor current revival. 'To a considerable degree, the change of ownership of specimens appears to have been effected under the immediate influence of wars and revolutions which shook the foundations of social order, or in connection with such violent events, even if ultimately in the form of legal purchase' (Wittlin). Dance (1966) gives examples of price deterioration in shells during the 18th century, which he was tempted to correlate with the rise of a scientific understanding of the subject, and with the early scientists' contempt for mere collectors. Despoliation brought much material as booty to France, the most famous fossil being the Maastricht *Mosasaurus hoffmanni,* brought to Paris in 1795. Many of such objects were returned after Napoleon's downfall but others found their way to sales in England. The breakdown of social barriers brought about by the French Revolution opened the doors to many palaces on the continent. Britain represented a stable bulwark (a fact which has been used to explain Lyell's uniformitarianism versus Cuvier's catastrophism), to which collections were transmitted for protection or safe disposal (Wittlin). Several of the catalogues listed here may refer to collections in this category, but they have yet to be investigated. There has been almost no study of 18th-century geology, and even less of the long history of fossil collecting: fossils have yet to have their Dance!

There has been a tendency to disdain the 19th-century attitude exemplified by the would-be shell collector who wished to procure a collection 'to satisfy his desire of knowing something of objects so varied and beautiful....then shall I have the gratification of knowing that I am better acquainted with subjects which have occupied the greatest minds and am more fit to associate with men who have won the admiration and esteem of their fellow creatures' (Matheson 1964). But such collectors probably formed part of a group within the 18th- and 19th-century scientific community that patronised science (less formally than through publication or education) by 'technical curiosity, cultural identification, rational amusement or theological edification' (Shapin & Thackray 1974). For the historian of science, 'science as people think of it and as they use it is every bit as important as science as scientists conceive of it'. A study of sale catalogues, particularly of house sales that list total house contents (e.g. Mantell 1853 and Lynch 1860), enables the earlier collector himself to be reconstructed, in the case of Balston (1919) even to his cigars. Rudwick (1972) has given a welcome insight into the history of palaeontology, but the prosography is to come. Investigation has yet to be made of the common background characteristics of fossil collectors 'by means of collective study of their lives, in order to tie exciting developments in intellectual and scientific history down to social, economic and political bedrock' (Shapin & Thackray 1974). For this task the material that follows provides a challenge.

REFERENCES

Allingham, E. G. 1924. *A Romance of the Rostrum*. London.
de Beer, G. R. 1953. *Sir Hans Sloane and the British Museum*. London.
Cleevely, R. J. 1974. The Sowerbys, the Mineral Conchology, and their fossil collection. *J. Soc. Biblphy nat. Hist.* **6**, 418-81.
Curwen, E. C. 1940. *The Journal of Gideon Mantell*. London.
Dance, S. P. 1966. *Shell Collecting, an Illustrated History*. London.
Eyles, J. M. 1967. William Smith: the sale of his geological collection to the British Museum. *Annls Sci.* **23**, 177-212.
Eyles, V. A. 1971. John Woodward (1665-1728): a bio-bibliographical account of his life and work. *J. Soc. Biblphy nat. Hist.* **5**, 399-427.
Jahn, M. E. 1963. *The Lying Stones of Dr. J. B. A. Beringer*. California
Jahn, M. E. 1966. The Old Ashmolean Museum and the Lhwyd collections. *J. Soc. Biblphy nat. Hist.* **4**, 244-8.
Laskey, J. 1813. *A general account of the Hunterian Museum, Glasgow*. Glasgow.
Matheson, C. 1964. G. B. Sowerby the first and his correspondents. *J. Soc. Biblphy nat. Hist.* **4**, 214-25.
Murray, D. 1905. *Museums, their history and their use*. Glasgow.
Owen, R. 1862. *On a National Museum of Natural History*. London.
Rudwick, M. J. S. 1972. *The Meaning of Fossils*. London.
Shapin, S. & Thackray, A. 1974. Prosography as a research tool in the history of science: the British scientific community 1700-1900. *Hist. Sci.* **12**, 1-28.
Sherborn, C. D. 1940. *Where is the – Collection?* Cambridge.
Smith, W. J. 1960. A Museum for a Guinea. *Country Life*, 494-5.

Wittlin, A. S. 1949. *The Museum, its History and its Tasks in Education.* London.

Woodward, A. S. 1904. *In* Lankester, E. R. (ed.) *History of the collections contained in the Natural History Departments of the British Museum.* Vol. 1. London.

MINERALS

PETER G. EMBREY*

Mineral specimens are natural history objects, but as a class they sit rather uncomfortably to one side and in some respects have affinities with specimens in the art and archaeology museums. Gemstones, for example, although belonging squarely in the field of minerals in their rough state, become objects of art when they have been carved or faceted by the lapidary, the more particularly when they have been set in an article of jewellery. The price of many mineral species tends to be controlled less by rarity or perfection of crystalline form than by their desirability as raw material for the craftsman to work on. At the other extreme a group of quartz crystals, one of the commonest of mineral species, may have an aesthetic disposition that gives it a cash value far and away beyond that occasioned by purely mineralogical considerations.

What is a mineral specimen? Like the other question, what is a mineral species, it has no simple answer because it involves subjective as well as objective considerations. A rock is an aggregate of one or more mineral species, and is the overall generic term; it becomes an ornamental stone, or an ore specimen, or a mineral specimen by entirely arbitrary decisions based on aesthetics, economics, or an interest in some aspect or other of one or more of the constituent minerals. A piece of granite may interest the petrologist by virtue of its origin, the monumental mason by its texture and colour, and the mineralogist by the shape of its feldspar crystals or the presence of traces of a rare species in an insignificant cavity.

To make broad generalisations is to invite instant contradiction on points of detail, but it may be fairly safe to say that biological material tends to be 'standardised' in a way that minerals are not. A crystal of spodumene may be anything from a couple of micrometres to fifteen metres long; it may be equidimensional, or long and prismatic; it may be colourless, or yellow, green, or pink; it may be transparent, or opaque because of alteration or inclusions of other minerals; and in specimen form it may be alone or associated with any combination of a dozen or more other mineral species. Now it may be said that these variations are entirely superficial and trivial, but they are nevertheless gross; one may consider a tawny owl, or a golden cowry, or a scarab beetle, to be a 'typical specimen' in a sense that is quite impossible for a crystal of spodumene. One may go a stage further and venture the statement that the essence of a fine mineral specimen is that it should be atypical; in nature adverse conditions for the growth of mineral crystals are the norm rather than the exception, and the well-developed crystal of almost any species is very unusual. It must be emphasised, however, that fine mineral specimens are not freaks in the sense that defective postage stamps are; it is simply that the average example of a mineral has been so badly deprived of nutrition and room to develop during

*Department of Mineralogy, British Museum (Natural History), Cromwell Road, London SW7

growth that it shows but few of the characteristics that it might. 'In the country of the blind, the one-eyed man is king.'

I hope that I have now gone some way towards establishing the idea that mineral specimens have a very real individuality, and this brings us to difficulties in the identification of mineral species. Compared with the entomologist who has to face some half a million species, the mineralogist may be thought to have an easy time with only two thousand or so; but the value of the accurate drawing or photograph is much greater to the biologist. A picture of, say, a Camberwell Beauty might be of any specimen of that butterfly, whereas an accurate illustration of a mineral refers to a particular specimen and only relates to other specimens by an exercise of the imagination. Minerals can certainly be identified, a majority without too much trouble, but the art of mineral illustration has never been of great help; per contra, the identification of the original of an illustration is made much easier, and it is a great pity that catalogues of mineral collections have so rarely contained artwork. The figured specimen, so important in connection with the designation of types in the biological sciences, has been relatively insignificant in mineralogy where characterisation has depended on - and contributed much to the development of - chemical analysis and the determination of the physical properties of crystals.

We have seen that the general run of minerals in their usual state are of little interest to the collector. Few but the most resistant minerals, such as gold and some of the gem species, are to be found in an attractive state at the earth's surface after millenia of exposure to the elements. All the good specimens have been exposed to view shortly before collection, either by earth movements or by mining and quarrying operations; the same is true to a large extent of palaeontological specimens, but of practically no other biological material. They find their way into the hands of local people, mainly the miners or quarrymen or - in Switzerland - mountain guides, and then usually through dealers into private or public collections. A successful collecting expedition for minerals that does not rely heavily on commercial rock-breaking and earth-moving activities is practically unknown. The other feature of minerals that should be noted is that, although they are localised by geological factors. they are not necessarily restricted geographically as are biological populations and may thus occur in any country where the general geology is favourable. In common with all specimen material, the one essential piece of information that cannot be determined unequivocally in the laboratory is the locality, and the absence of this is paramount in reducing the scientific value of a specimen; far too few collectors, both in the past and at the present day, have paid much attention to this and nearly all collections contain a depressingly large number of vague or conjectural localities. Another discouraging matter, of particular relevance in this essentially historical work, is the prevalent lack of interest in documentation; when specimens have passed through a number of hands it is rare indeed to be able to trace them back by more than a single step. Outstandingly exceptional in this respect are several specimens sold by the dealer J. H. Heuland (1778-1856) to Louisa, Countess of Aylesford (1760-1832); on her death, Heuland sold her collection in the years 1834-1839 and one of the buyers was Isaac Walker (1794-1853); in 1911 the Walker collection was bought by the dealer Samuel Henson (1848-1930), and some specimens were bought by the British Museum (Natural History) in 1912; other specimens were bought by F. N. A. Fleischmann (F. N. Ashcroft; 1878-1949) and by (Sir) Arthur Russell (1878-1964), and these later came to the Museum by donation or bequest. Entries for these specimens can be found in

Heuland's sale catalogues (twice) and in Lady Aylesford's catalogue, and I know of no others so well travelled and documented; the only information missing is Heuland's original source!

Cut gemstones and other examples of the lapidary art are known from very considerable antiquity, but there are few mineral specimens as such with pedigrees earlier than the 18th century. There can be no doubt whatever that such authorities as Pliny, Aristotle, and Theophrastus in classical times owned minerals, but their collections are irretrievably lost; even Agricola (1494-1555) left only his writings to posterity. There must, likewise, have been many excellent specimens collected by monarchs and noblemen down the ages but these too have mainly found their way into national museums or have been lost. At the present time, Chatsworth House is almost the only stately home with mineral specimens, and many of these, so I understand, are recent purchases by the present Duke of Devonshire. There are at least two reasons for this unhappy state of affairs: unless they are kept in cabinets, the specimens are too easily separated from their labels - if there ever were any; and although 'stones', mineral specimens are readily chipped and broken if handled carelessly, dust and dirt accumulate to hide their beauty, and the result looks so much like a worthless lump of rock that it is thrown away. It takes only a few bruises to reduce a prime specimen by 90% or more of its cash value, a fact unknown to heirs who pile a collection carelessly into a crate or leave the best pieces lying around on the mantelpiece or a coffee-table and then feel they are being cheated when they try to sell.

Auction sales of mineral specimens, as the present list shows, have been taking place for the last 250 years. Significant sales, however, have been the exception; in other fields of natural history type specimens of species have frequently changed hands at auction, but I know of no mineral type specimen in this category. There is no doubt, as the collection of the British Museum (Natural History) can testify, that fine cabinet and display specimens have been acquired at auctions, but scientific interest has nearly always resulted from investigations carried out after purchase. One of the rare exceptions was the auction at Leith in 1808 of Giesecke's specimens from Greenland, at which Thomas Allan and Colonel Imrie recognised cryolite, the first specimens to be seen in Britain. There were, of course, sales where the collection as a whole was of scientific interest. The collection of the crystallographer and mineralogist the Abbé R. J. Haüy (1743-1822) was bought from his executors by the 1st Duke of Buckingham and Chandos for £4000; at the house sale in 1848 of the contents of the Duke's seat at Stowe, Buckinghamshire, it was bought as a single lot for £325-10-0 by the Directors of the Jardin des Plantes, Paris, but it had apparently suffered considerably from the decomposition of many specimens.

When one looks at the sale catalogues it is difficult at this distance in time to distinguish between quality and empty eulogy, and few of the named buyers (where these are recorded) ring any bells as being the owners of notable collections. The reasons for this are entirely conjectural: were the specimens bought to add to a general collection of natural curiosities, later dissipated; were they bought simply as decorative objects, as many are at the present day; or did the buyer tire of them and sell to a dealer during his lifetime? We may surmise that mineral collectors have been more secretive about their affairs than have collectors in other fields, but this seems improbable if only because it asks the question Why. We may also ask why mineral auctions have been relatively unpopular, and here, I think, there is a reasonable answer. Mineral collectors of taste and means, even in the heyday of collecting, early in the 19th century and again in the last decade or so, have been relatively few in number, and the same may be said of knowledgeable and enterprising dealers. At any time in the last two centuries there have been some of the latter who would travel extensively in search of new sources of fine specimens, and who would take the trouble to find out the tastes and needs of their best customers so that they were never at a loss to place a good specimen at the right price. Such dealers have been specialists in their own right, and have let the poorer specimens pass down to less discerning or less active dealers. This is, of course, true in any field of collecting but mineralogy has always been a minority taste. The big auction houses have been able to pay specialist members of staff good salaries in fields of majority interest where the turnover is large and profitable, but natural history objects and minerals in particular have usually constituted a sideline. From the customer's point of view it has always been more satisfactory to be able to get specimens on approval and examine or verify them at leisure, rather than have to make a snap decision at a preview.

[John] Henry Heuland, whom I have mentioned above, will long remain the shining exception to the general rule, for here was a specialist dealer who for many years conducted the bulk of his business by auction. Clearly it suited his taste to collect specimens on his travels, and then 'by a cleverly-arranged series of public sales [he] seems to have done much towards keeping alive the interest and rivalry of collectors'. Little is known of Heuland's life apart from his activities as a dealer, which occupied the years from 1808 (or perhaps a few years earlier) until 1848; it is probable that the last years of his life were inactive since he died of 'softening of the brain' in 1856. Heuland's uncle was [Adolarius] Jacob Forster (1739-1806), an active dealer who travelled extensively in pursuit of specimens leaving - for at least some of the time - his wife, Elizabeth, in charge of their shop in Gerrard Street, Soho. Elizabeth's brother, George Humphrey, was also a dealer in minerals and other natural history specimens. Forster had another shop in Paris, managed by his brother Ingham [Henry] Forster, and held sales there in 1760, 1769 and 1783. In his will, Forster left his private collection and most of his stock jointly to his brother and to Heuland, the latter to sell them (after cataloguing the private collection) and to divide the proceeds. The private collection, after much delay, was catalogued by Armand Lévy and sold to C. H. Turner, from whom it passed to Henry Ludlam and is now in the museum of the Institute of Geological Sciences, London. With the exception of two by Christie's in 1812, all of Heuland's sales were held at 38 King Street, Covent Garden, the premises of King and Lochée and their successors (King; Dubois; Thomas; Thomas & Stevens; and finally Stevens). He held some forty or so sales, the exact number depending on one's definition of what constitutes an individual sale. Although Heuland's name is not mentioned on the catalogue, the terms of Forster's will make it certain that he was responsible for the sale of Forster's stock in 1808; this must be the largest mineral auction on record, having taken place in five parts over a total of forty-one days and including 5,860 lots. The catalogue runs to 458 pages, and the sole extant (to my knowledge) copy, in the British Museum (Natural History), contains prices but no names of buyers. The composition of sale catalogues is the responsibility of the auction house, but Heuland's sales seem to be unusual in that he compiled all his own entries.

The descriptive language used in catalogues of a century and more ago makes fascinating reading, and a detailed study of it would make an interesting project.

There are some differences in spelling but even greater similarities in style between G. B. Sowerby's catalogue (1827) of Sir Alexander Crichton's collection and contemporary Heuland catalogues; whether there was conscious influence, or whether Heuland assisted Sowerby - or even had a hand in Crichton's own cataloguing - we shall probably never know. I may mention here that, although the Crichton sale catalogue is not annotated with names, a few of the specimens are now known through their appearance at the sale of the Walker Collection, bearing Walker's notation *C-1827* on his very characteristic labels.

On looking through the remainder of the sales (how one grows to dislike ANON!) only three more catch the eye: John Hawkins (1905), John Ruskin (1971), and the Duke of Northumberland (1968). The collection of John Hawkins (1761-1841) was fairly complete, and contained many old Cornish specimens; (Sir) Arthur Russell was a buyer at this sale, and his collection is now in the British Museum (Natural History). The Northumberland specimens had been kept for many years in a turret, with much consequent damage and decay, and the collection as a whole was a poor remnant of its former quality. The Ruskin collection was one that he had given in 1883 to St. David's School, Reigate, later at Stone House, Broadstairs, Kent; at different times he had given educational collections to some nine separate establishments.

We can, thus, see that, with the exception of Heuland's activities, sales by auction have constituted an infinitesimal part of the overall trading in mineral specimens, and I have essayed one or two suggestions as to why this should have been the case. In the past nearly all the notable collections have tended to change hands as single units, by sale, gift, or bequest, rather than broken up by auction. The Haüy collection was sold as a single lot (see above), and at the same sale the Duke's own collection (the Buckingham or Stowe Collection) also went as a single lot. The 6000 specimens were bought for £68-5-0 by the dealer James Tennant, who must have taken out and sold many of the best specimens before cataloguing and offering for sale by 1865 a collection 'founded on the Stowe.....collection .....' [ms catalogue with 1937 entries, BM(NH)]. This collection was eventually bought about 1877 for £3000 by John Ruskin, who then (after advice) considered that he had been swindled and sued Tennant. Ruskin settled out of court, and after legal fees of £1000 had been paid received a further £1000 and specimens to the value of £500. An illustration of the more usual progression of significant private collections of minerals is given by listing the present locations of those consulted by R. P. Greg and W. G. Lettsom when they compiled their Manual of the mineralogy of Great Britain and Ireland (1858): C. H. Turner's and W. Nevill's collections were acquired by Henry Ludlam* and bequeathed to the Museum of Practical Geology (now the museum of the Institute of Geological Sciences) in 1880; the Allan-Greg collection was bought by the Trustees of the British Museum in 1860 from R. P. Greg (1826-1908); H. J. Brooke's collection, on his death in 1857, passed to Cambridge University; M. F. Heddle's collection is in the Royal Scottish Museum, Edinburgh; and T. Thomson's collection is in the Hunterian Museum, Glasgow.

At the present time there seems to be little chance of significant mineral auctions, except perhaps in the field of material suitable for interior decoration. There has been a considerable growth of interest in mineral collecting in the last decade or so, and there is a shortage of fine specimens to satisfy the many weal-

---

*A magnificent specimen or proustite, from Chañarcillo, Chile, was bought for £200 by Henry Ludlam at Stevens' sale rooms on 28 November 1876, and presented to the British Museum (B.M. 50811).

thy collectors. Modern mining methods contribute to the shortage of specimen material because they concentrate on vast tonnages rather than small rich deposits, and even in the latter the explosives in current use shatter crystals more than the older blasting powders. Miners are more aware than ever before of the value of the specimens, and despite controls by the mine-owners can often hide in their clothing the equivalent of one or two months' pay. In these conditions many of them will skilfully play off one dealer against another, and so the prices rise.

What constitutes a high-value specimen? We have seen already that aesthetic appeal generally counts for more than rarity, although a combination of the two is unbeatable, and that there is a heavy premium on perfection of crystallisation. Sir John Pope-Hennessy recently said 'There is no such thing as taste, there is only quality': an interesting and provocative subject for debate, but taste and fashion seem in practice to determine the correlation between quality and price. But what determines fashion I have never been able to discover. One can only observe the market.

REFERENCES

Frondel, C. 1972. *Mineral. Mag.* **38**:545 (Jacob Forster)
Russell, A. 1950. *Mineral. Mag.* **29**:395 (Henry Heuland)
Spencer, L. J. 1912. *Mineral. Mag.* **16**:251 (Isaac Walker)
Sweet, J. M. 1974. *Annals of Science* **31**:25 (C. L. Giesecke)
Whitehead, P. J. P. 1973. *Mineral. Mag.* **39**:361 (Jacob Forster)

SHELLS

S. PETER DANCE*

Molluscan shells have been offered for sale by auction since the late 17th-century. Unlike insects, mammal and bird skins, birds' eggs, pressed plants and fungi, they are relatively indestructible and in this respect are more akin to minerals, fossils and bones. Their durability, as much as their beauty, explains the early popularity of shells in European auction houses. Less durable natural objects require troublesome preparation for the cabinet, are prey to insects and mould once in it, and out of it they cannot long withstand damage. The greater popularity of shells and similar objects at auction continued until sophisticated methods of preparation and efficient storage of the more vulnerable natural objects were introduced. Consequently the history of shells at auction is impressively long and continuous. It reaches a high point in the second half of the 19th century and declines slowly thereafter. It is too soon to evaluate the revival of interest that has taken place since the 1960's.

Collectors who like shells usually like other hard-wearing natural objects too, the durability and solidity of shells, barnacles, minerals, fossils and bones giving them a unity, in a cabinet context, which their dissimilar appearance denies them. This explains why shells kept mixed company in cabinets up to recent times. Even now the presence in a shell collection of barnacles, sea urchins and pebbles is not considered unusual. It also explains why there have been so few sales devoted wholly to shells and why collective sales of shells, fossils and minerals still take place.

The earliest recorded auctions of shells and related 'curiosities' took place in 17th-century Holland where, for many years, they were very popular. E. F. Gersaint, friend of the painters Boucher and Watteau, visited Holland in the 18th century and noticed the great interest shown in shell auctions. Subsequently he conducted many Paris auctions of natural and artificial objects, among which shells were a principal attraction. Inevitably the fashion reached London. By the 1750's sales of natural objects were fairly frequent, but the first sale worth mentioning from a conchological standpoint was that of Commodore Lisle's collection in 1753. On that occasion four specimens of the Precious Wentletrap (*Epitonium scalare L.*) brought very high prices, one of them realising £23. 2s.

During the greater part of the 18th century the conchological centre of the world was Holland, its trade with the East and West Indies providing many opportunities for its sailors and traders to acquire exotic shells. The three circumnavigations of the globe by Captain Cook's vessels were instrumental in shattering the Dutch monopoly in shells and many other commodities. From 1771 onwards European shell collections were enriched by specimens acquired by British voyagers and often the shells passed through a London auction house before ending up in a collector's cabinet. Among the earliest auctions to offer

*Priory Hall, Priory Wood, Clifford, Herefordshire

shells obtained during one of Cook's voyages was one held at Langford's in November 1775. Among the many items from the South Seas were shells previously unknown in Europe including the Golden Cowry (*Cypraea aurantium* Gmelin) and the Sun Shell (*Astraea heliotropium* Martyn). The University Museum of Zoology at Cambridge has a documented Sun Shell which is known to have been sold then. As far as is known this is the earliest documented example of a shell obtained at auction in Britain extant.

The first outstanding shell collection to be auctioned in Britain was that formed by the Duchess of Portland. In 1786 her huge museum of natural and artificial objects was sold in a vain attempt to meet the demands of her creditors. Unlike most collections of that era the natural-history portion was in very good order having been curated by Daniel Carl Solander, the favourite pupil of Linnaeus. As the Portland shells were the finest available at that time there was strong competition for them at the sale; many years later some of them were still bringing high prices at auction. The catalogue advertising the Portland sale was an impressive publication and was to become of lasting importance, once it was realized that it utilised valid binominal nomenclature to denote most of the zoological specimens zoological specimens that many scientifically correct names were first published therein. Another unusual feature of the catalogue is the frequent appearance of locality information in the descriptions of lots. Compared with earlier -and many later - sale catalogues the entries have an authoritative and scientific appearance:

> Lot 1603. Venus erosa, *S*. [*Batissa triquetra* Deshayes] a large and singular fresh-water Bivalve, from New South Wales, extremely rare.
> Lot 3831. An exceeding fine and large Cypraea aurora, *S*. or the Orange Cowry [*Cypraea aurantium* Gmelin], from the Friendly Isles, in the South-Seas, extremely scarce.
> Lot 3906. Nautilus scrobiculatus, *S*. or Great umbilicated Nautilus [*Nautilus scrobiculatus* Lightfoot], from *New Guinea, very rare*.

Unfortunately the scientific tone of the Portland sale catalogue had little cash value and was not maintained in subsequent catalogues.

Just how uninformative and unscientific a sale catalogue could be in those days is adequately shown by the following entries in one advertising shells and other natural objects sold by Mr King at Covent Garden from 22 to 25 April 1799:

> 1st day, Lot 102. Two large Arabian spindles, and a scarce alatus shell.
> 2nd day, Lot 41. Two small spiders, 2 young ditto, and a young devil's claw shell.
> 4th day, Lot 108. A pomegranet sea snail, finely variegated by partially taking off the skin, a camp olive [*Oliva porphyria* L.], and a sardonyx cowry.

It was twenty years before a sale comparable with that of the Duchess of Portland's was held. The sale of Sir Ashton Lever's collections in 1806, long after they had passed out of his ownership, attracted many collectors and dealers and brought on to the market many objects of considerable value and rarity. Many of the shells originated from the Cook voyages and some had been acquired at the Portland sale. The first quarter of the 19th century produced few good shell sales, the Napoleonic wars effectively inhibiting collecting and selling. The collection formed by the eccentric Henry Constantine Jennings, which was auctioned in 1816 and 1818, must have been outstanding, however, because it was once valued by the British Museum authorities at 9,000 guineas and its owner is known to have paid very high prices for specimens. An annotated sale catalogue could indicate where some of his treasures went.

The Bligh sale of 1822 is very well known by comparison. The published catalogue - itself a collector's item - was largely written by William Swainson, who described many species in it and embellished it with two excellent coloured plates. C. Dubois, who auctioned the collection, was aware that a well-presented catalogue could help to push the bids up and the Bligh sale was a notable success in commercial terms. The sale of Mrs Bligh's shells (many of them collected by her husband, the unfortunate William Bligh of the *Bounty*, during his voyages to the South Seas) marks the end of an era dominated by the constant re-appearance at auction of material originally collected by men associated with the voyages of Captain Cook. After a few years when little of conchological importance happened in the salerooms, a new era began. It was not to end until the advent of the Second World War.

The activities of the redoubtable collector Hugh Cuming provided the impetus for the new era. Through his shell-collecting expeditions among the islands of Polynesia, along the west coast of South America and throughout the Philippine Islands Cuming had amassed the largest and most valuable collection of shells ever brought together by one person. Returning from the Philippines in June 1840, he amazed the conchological world with his seemingly inexhaustible stores of new and beautiful specimens. He was quick to profit from his labours by the sale of duplicates and it soon became almost impossible to build up a collection of high quality specimens without purchasing from him.

It became quite impossible to write up-to-date descriptive works on conchology without access to the Cuming Collection. George Brettingham Sowerby (both first and second of the name) based their *Thesaurus Conchyliorum* on his specimens; and Lovell Reeve's *Conchologia Iconica*, the most complete and most sumptuous of all shell iconographies, became virtually an illustrated guide to the Cuming Collection. Any shell on which a drawing in the *Conchologia Iconica* was based was certain to merit a glowing reference in a sale catalogue and to attract spirited bidding in the auction room.

The sale of John Adamson's shells in February 1848 was the first of many which showed Cuming's beneficial influence. This was a modest sale, 720 lots being auctioned for a total of £84.3s, but it was not a failure by contemporary standards. The A. L. Gubba sale of January 1851 was more interesting because it contained many rarities and the catalogue made more compelling reading. A liberal use of italic and upper-case type faces helped to distinguish the better quality material from the more mundane, the cataloguer having an obviously sympathetic understanding of the collectors' interest. Entries such as 'CYPRAEA TESSELLATA, *extremely fine and scarce*', and 'MITRA REGINA AND SANGUISUGA, *both large and beautifully marked specimens*' are typical of the eye-catching style which came to be characteristic of catalogues advertising the sales of well-endowed collections.

Many good shell collections were being built up in the 1850's but few were sold then. The only other important sale was that of H. Vernède's collection in December 1859. This collection contained outstanding treasures such as The Glory of the Sea (*Conus gloriamaris* Chemnitz), *Conus thomae* Gmelin, and two specimens, the only ones known, of *Ancilla vernedei* Sowerby, the finest and certainly the rarest species of the genus *Ancilla*. Lovell Reeve's collection, sold in May 1864, also included many interesting and scientifically valuable shells. In the following year, however, these, and most other British shell auctions were eclipsed by the sale of the incomparable collection formed over a long period by John Dennison.

From 24 to 29 April 1865 J. C. Stevens auctioned 'The celebrated Dennison collection of shells ... being one of the finest ever offered to public competition'. The elaborate catalogue was drawn up by the Liverpool conchologist Frederick Price Marrat who identified the specimens and arranged the lots. Marrat's task must have been formidable. Dennison had a passion for beautiful and rare shells but he was no lover of order and cared nothing for science. Indeed, Reeve noted in his diary in 1849 that Dennison stored his shells in a costly rosewood cabinet decorated with intricate carving and painted panels, but the shells were placed in the velvet-lined drawers 'like a mess of porridge, without names ... By paying liberal prices for specimens all the rarest and best that come into the country are offered him.' The catalogue shows that Dennison had an eye for superlative shells and evidently a purse deep enough to satisfy his desire to own them.

An arresting feature of the collection was the large number of specimens which were the originals of illustrations in the *Conchologia Iconica,* the illustrious connection of Dennison's superb shells with Reeve's exquisite pictures being emphasised throughout the catalogue.

> Lot 551. CONUS CEDO-NULLI (THE SHELL FIGURED 46E., CONCH. ICON.) *A dark jet purple shell of surpassing rarity and beauty, to which neither figure nor description can do justice* - vide *Reeve.*
> Lot 735. VOLUTA RETICULATA, BEAUTIFULLY PENCILLED, VERY RARE, SPECIMEN FIGURED IN CONCH. ICON.
> Lot 771. ONISCIA EXQUISITA, SPECIMEN, figured in *Conch.* Icon., a highly interesting shell, and of the utmost rarity.

No previous shell sale had included so many rarities, any one of the following list being sufficient to make a collection outstanding: *Cypraea guttata* Gmelin, *Cypraea valentia* Perry, *Conus gloriamaris* Chemnitz, *Conus thomae* Gmelin, *Conus cervus* Lamarck, *Festilyria festiva* Lamarck, *Morum dennisoni* Reeve, *Mitra dennisoni* Reeve, *Pholadomya candida* Sowerby. It is unlikely that such a glittering array of conchological treasures will ever be offered again at a single sale.

Most of the big names in the British conchological fraternity attended the sale: Reeve, the Sowerbys, G. F. Angas, Barclay, Lombe Taylor, the young J. C. Melvill then at the start of his long career as a collector, and Cuming, doyen of them all, whose career, like Reeve's was to end a few months later. They were compelled to come, for this was the shell sale of the century. It is ironic that the public dispersal of a collection which for so long had resembled 'a mess of porridge' should become the apogee of the Victorian shell craze. At the same time it is saddening to realize that the innocent enjoyment of a shell's intrinsic beauty, fervently manifested by those who attended the Dennison sale, has since been diminished, though certainly not killed, by the clamorous demands of scientific rectitude.

In May 1866 a very different sale was conducted by Mr. Stevens. For the first time in England a collection was sold originating almost entirely from one limited geographical area: Mauritius and its satellite islands. It was pedestrian compared with Dennison's but included several rarities which had eluded that veteran collector. The two most noteworthy of these were:

> Lot 93. CONUS BARTHELEMYI, ONE OF THE HANDSOMEST AND SCARCEST OF THE CONES, THERE BEING ONLY ONE OTHER SPECIMEN KNOWN.
> Lot 120. CYPRAEA BRODERIPII, A MOST BEAUTIFUL SHELL, THE GEM OF THE COLLECTION, ONLY THREE OTHERS KNOWN.

It was nearly a hundred years later that the *Cypraea,* almost as scarce now as it was then, was recorded from a precise locality and later still before it was authenticated for Mauritian waters. The *Conus* is still scarce and only recently has it become available again from Mauritius. Here is a rare instance of a sale catalogue providing information adumbrating much later discoveries.

The year 1873 witnessed the demise of three excellent collections. That of Sigismund Rucker, sold in March, was particularly rich in rare species of *Conus* and large showy shells, such as *Spondylus*. The Rucker sale was followed, in June, by that of the first portion of Thomas Norris's collection. Again Marrat was the cataloguer and in a prefatory note to the catalogue said: 'Mr. Norris and Mr. Dennison were rival collectors during a number of years, and it was a disputed point, not only between these two gentlemen, but between the best conchologists of their day, as to which had the best collection.' Indisputably Dennison's was more replete with rarities, but in some respects Norris had the edge on his rival. In particular his collections of *Conus* and *Mitra* were more extensive. Of the latter group Marrat said: 'his collection became very famous, from such a large number of specimens being figured in Reeve's *Conchologia Iconica*'. Marrat considered the collection to be 'among the last of the great collections that will be offered for public competition'. The second portion was auctioned in July.

The other notable sale of 1873 was similar to that of May 1866. Again the collection originated from Mauritian waters and again the owner remained anonymous (though one may hazard a guess that he was Victor de Robillard or, less likely, Sir David William Barclay). This time the catalogue is refreshingly clear in its reference to localities. Typical of the catalogue entries are the following:

> Lot 156. Pretty specimen of Conus *timorensis* and the rare C. CALEDONICUS, of Reeve, both from St. Brandon Shoal [Cargados Carajos].
> Lot 168. Three *Daphnella,* new species; 2 *Clathurella Robillardi,* with specimen of the rare *Rissoina insolita* of Deshayes, all from Barkly Island.

Whether or not the locality information helped to raise the bids is not known.

Of the remaining Victorian shell sales three deserve special mention. In June 1880 the wonderful collection of Thomas Lombe Taylor was auctioned. It is known to have cost its owner £17,000 to accumulate and is more noteworthy for what was *not* included in the sale because the British Museum (Natural History) had been presented with many of the gems beforehand, and G. B. Sowerby (the third of the name) purchased many others before the sale. In addition a large residual collection was retained by Taylor's son and this was not auctioned until 1929. Few natural history collections can have taken so long to be dispersed.

The Harford collection, auctioned in July 1884, was another major conchological event. A note in the sale catalogue states: 'Whatever the Proprietor bought was the best available, and most of the Shells now sold were purchased at the Dennison, Rucher [sic!], Vernede and Anges Sales, and at the dispersion of all the important collections which have been sold in Mr. STEVENS' Rooms during the last Forty Years.' There were many valuable shells in the collection, including some illustrated in the *Conchologia Iconica,* but the most outstanding items were not shells at all. Most late-19th-century shell auctions concluded with the sale of the cabinets and the library of conchological books. Harford's cabinets and books were the highlights of his sale. Most collectors now would prefer Lot 632, a subscriber's copy of the *Conchologia Iconica* complete in twenty volumes, to the entire Harford shell collection: it is doubtful if they would have shown the same preference in 1884.

The last major shell sale of the 19th century was also one of the most important. The collection formed by Sir David William Barclay was put up to auction in July 1891, the catalogue having been prepared by Hugh Fulton, then at the beginning of his career as a shell dealer. It is unlikely that any of the numerous collections Fulton handled in later years approached Barclay's in value and interest, for it contained many type specimens as well as several extremely rare items. The specimen of *Strombus taurus* Reeve was to be for many years the only one known. Curiously the prices obtained at the sale were very low considering the high quality and rarity of the shells offered. One reason for this may be the presence in the saleroom of the wealthy collector J. J. MacAndrew who bought many of the choicest shells. A passage in an unpublished letter from G. B. Sowerby (3rd) to J. C. Melvill suggests that MacAndrew's presence was an inhibiting factor: 'Poor Mr. Higgins did not get the *Marginella mirabilis* after all, he asked me to bid up to £5 for it for him, it sold for £6. 10. 0! The *Strombus taurus* went for £5. 10. 0. that would have gone higher but I would not bid against Mr. MacAndrew' (MS in National Museum of Wales, dated 14 July 1891). The two shells mentioned were bought by MacAndrew (and acquired subsequently by Melvill).

Despite the huge stocks of land shells brought into the country by Cuming, shell sales had always been overwhelmingly marine in character. The situation now changed because some dealers, notably Fulton and H. B. Preston, began to specialise in land and freshwater shells. The minor sale of G. B. Mainwaring's collection in August 1894 offered an almost equal number of lots devoted to marine and non-marine species. The sale of T. Vernon Wollaston's Madeiran land and freshwater shells in January 1900 was followed, in May 1904, by the sale of Hugh Nevill's collection, rich in the land and freshwater shells of Ceylon, and by that of J. C. Cox's collection in July 1904 (and February 1905), containing marine and non-marine shells in about equal proportions. One of the leading collectors of land shells was Solomon J. Da Costa whose magnificent collection consisted largely of specimens collected in South America and elsewhere by collectors employed by him for that purpose. At the sale of his shells in October 1907 most of the acknowledged rarities were displayed including 'a magnificent specimen' of *Sultana labeo* Broderip (Lot 242) often regarded as the greatest prize among land shells. A large number of lots included paratypes of species described by Da Costa. The 'Golden Age' of the British shell auction ended, in February 1913, with the sale of Carl Bülow's large collection which was also largely made up of land shells. The sale was notable for the low prices paid, a mere nine shillings sufficing to secure Lot 157 'containing an extensive collection of Pomatias, in over 100 glass-topped boxes'. Since then land and freshwater shells have never been widely popular, in or out of the saleroom.

After the First World War shell auctions became rare events. Sales of insects, by contrast, gained in popularity. Whether shell sales diminished because of waning interest or whether interest waned because of the dearth of sales is not clear. Paradoxically the Second World War helped to revive interest in exotic shells. Troops stationed overseas often had the opportunity to look for shells in tropical and sub-tropical waters and brought them back home. A new enthusiasm for shell collecting was kindled and a new era initiated, not until 1969 did shells of high quality re-appear for sale in a leading London auction house. that of Sotheby & Co. Shells as objects to collect for pleasure, and hopefully as objects of investment value, are now, along with minerals and certain fossils, more popular in the saleroom than insects. The more durable natural objects have now, as before, become more popular in the saleroom than the less durable ones. Once again the accent is on rarity and already some very high prices have been given for desirable species, particularly for those belonging to favoured genera such as *Conus* and *Cypraea*. Shells are again in great demand, though less so in Britain than in several other countries. Auction houses are playing a part in helping to satisfy that demand. That they may play as vital a part as they once did remains to be seen.

REFERENCES

With the exception of data taken from J. M. Chalmers-Hunt's *Register* and from sale catalogues all the information in this chapter may be found in the following works. The principal sources are my own *Shell Collecting* and *Rare Shells,* but much useful information is available only in the series of articles by Tomlin.

Allingham, E. G. 1924. *A Romance of the Rostrum.* Witherby, London (References to shell sales and prices obtained)

Cooke, A. H., Shipley, A. E. & Reed, F. R. C. 1895. *Molluscs and Brachiopods.* Cambridge Natural History Series, Vol. 3, Macmillan, London ('Prices given for shells', pp. 121-22)

Dance, S. P. 1966. *Shell Collecting: An Illustrated History.* Faber & Faber, University of California Press

Dance, S. P. 1969. *Rare Shells.* Faber & Faber, University of California Press

Dance, S. P. 1971. The Cook Voyages and Conchology, *J. Conch. London.,* **26**: 354-79 (Extracts from early sale catalogues)

Tomlin, J. R. le B. 1942. 1949 Shell Sales (I-VI), *Proc. malac. Soc. Lond.,* **24**: 157-60, **25**: 25-33, 96-99, **27**: 254-56

Wagner, R. J. L. & Abbott, R. T. 1967. *Van Nostrand's Standard Catalog of Shells* (2nd edition) D. Van Nostrand, Princeton ('A Brief History of Shell Auctions', pp. 7-9)

PART TWO

The Register of Natural History Sales

Signs and Abbreviations

add = additions
Anon = Anonymous
App = Apparatus
Bot = Botany, Botanical
Br = British
Butts = Butterflies
Cab = Cabinet(s)
Cat = Catalogue(s)
Col = Coleoptera
Coll = Collection
Cont = Continental
Cor = Coral(s)
Dipt = Diptera
Dup(s) = Duplicate(s)
f = few
ffdd = following days
For = Foreign
Fos = Fossil(s)
Hem = Hemiptera
Herb = Herbarium, herbaria
HH = Heads & Horns
Hym = Hymenoptera
inc = incomplete
incl = including

Ins = Insects
Irr = Irrelevant items
Lep = Lepidoptera
Lib = Library, Books
m = many
Mam = Mammals
Met = Meteorites
Min = Minerals
n = names of buyers
n.d. = no date
Neur = Neuroptera
Nh = Miscellaneous natural history
Od = Odonata
Ool = Oology
Orth = Orthoptera
p = prices
Pal = Palaearctic
Quad = Quadrupeds
Rel = Relevant
Rept = Reptiles
s = some
Sh = Shells
† = deceased

Catalogue Locations

AB	Dr. A. Blok, *in* the Library, Royal Scottish Museum, Edinburgh.
AR	Arbroath Public Library, Arbroath, Angus, Scotland.
ASH	The Library, Ashmolean Museum, Oxford.
B	Bodleian Library, Oxford.
BIRM	Local Studies Department, Reference Library, Birmingham Public Libraries, Birmingham.
BMETH	Department of Ethnography, British Museum, 6 Burlington Gardens, London.
BML	Reading Room and North Library, British Museum.
BMM	Department of Coins and Medals, British Museum.
BMP	Department of Prints and Drawings, British Museum.
BMNHB	Library, Department of Botany, British Museum (Natural History), South Kensington.
BMNHE	Library, Department of Entomology, British Museum (Natural History), South Kensington.
BMNHL	General Library, British Museum (Natural History), South Kensington.
BMNHM	Library, Department of Mineralogy, British Museum (Natural History), South Kensington.
BMNHP	Library, Department of Palaeontology, British Museum (Natural History), South Kensington.
BMNHZ	Library, Department of Zoology, British Museum (Natural History), South Kensington.
BRAD	Radcliffe Science Library, South Parks Road, Oxford.
BSMB	Bibliotek de Staatlichen Museen, 1-3 Bodestrasse, Berlin C2.
BUP	Bibliotheque d'Art et d'Archaeologie de L'Université, Paris.

C	Messrs. Christie and Manson, Auctioneers, London.
CAMB	Balfour Library, Department of Zoology, University of Cambridge, Downing Street, Cambridge.
CAMMUS	Museum, Department of Zoology, University of Cambridge, Downing Street, Cambridge.
CEB	Cabinet des Estampes, Staatlichen Museen, 1-3 Bodestrasse, Berlin, C2.
CEP	Cabinet des Estampes, Bibliotheque Nationale, Paris.
CH	J. M. Chalmers-Hunt
CI	Messrs. P. & D. Colnaghi, 14 Old Bond Street, London.
CLD	Library, Courtauld Institute of Art, 20 Portman Square, London.
DIP	Department des Imprimes, Bibliotheque Nationale, Paris.
EG	Edward Grey Institute of Field Ornithology, South Parks Road, Oxford.
EKW	Professor Sir Ellis Waterhouse, Paul Mellon Centre, London.
F	A. J. Finberg, *in* the Department of Prints and Drawings, British Museum.
FK	Frick Library of Art, New York.
FM	Fitzwilliam Museum, Cambridge
G	Bibliotheque de l'Université, Ghent, Belgium.
GEOL	Library, Geological Society, Piccadilly, London.
GLAS	Library, Hunterian Museum, University of Glasgow.
GS	E. Gowing-Scopes, Rosewood, Stonehouse Road, Halstead, Kent.
H	Library, Hope Department, University Museum, Oxford.
HEH	Henry E. Huntington Library, San Marino, California, U.S.A.
J	Messrs. R. B. and D. B. Janson, 44 Great Russell Street, London.
JAG	Dr J. A. Gibson, Foremount House, Kilbarchan, Renfrewshire, Scotland.
JR	John Rylands Library, Manchester.
LC	L. Christie, 137 Gleneldon Road, Streatham, London.
LINN	Library, Linnean Society, Piccadilly, London.
LNSW	Linnean Society of New South Wales, Sydney, Australia.
LIV	Local History Department, Liverpool City Library, Liverpool.
MAR	Rev. J. N. Marcon, Raydale, Fittleworth, Pulborough, Surrey.
M	Public Library, Middleton, Lancashire.
MDB	Martin Dawson Brown, The School House, Acaster Selby, York.
N	R. Nichols, 84 Old Farleigh Road, Selsdon, Surrey.
NG	Library, National Gallery, London.
NLS	National Library of Scotland, Edinburgh.
NMW	Library, Department of Zoology, National Museum of Wales, Cardiff.
NOR	Norwich Central Library, Bethel Street, Norwich.
NPG	Library, National Portrait Gallery, London.
NYK	New York Public Library, New York.
P	Messrs. Phillips Son and Neale, *in* The Wallace Collection, Manchester Square, London.
RCS	Library, Royal College of Surgeons, Lincoln's Inn Fields, London.
RES	Library, Royal Entomological Society of London, Queen's Gate, South Kensington.
RIA	Library, Royal Irish Academy, Dublin.
RKDH	Rijksbureau voor Kunsthistorische Documentatie, The Hague, Holland.
S	Messrs. Sotheby and Co., London.
SAR	Saruman Library (Paul E. Smart), Tunbridge Wells, Kent.
SPD	S. P. Dance, Priory Hall, Priory Wood, Clifford, Herefordshire.
SR	Seymour de Ricci, *in* Bibliotheque Nationale, Paris.
T	Alexander Turnbull Library, Wellington, New Zealand.
VA	Library, Victoria and Albert Museum, South Kensington.
VAE	Dr V. A. Eyles, Great Rissington, Glos.
W	Wallace Collection, Manchester Square, London.
WHB	W. H. Barrow, 1520 Melton Road, Queniborough, Leicestershire.

Explanatory Notes

COLUMN 1 is headed by the year or years of the period covered on each page. The dates are arranged chronologically as far as my information goes. But in the case of a sale of which only the year and month is known, and not the day or days when it was conducted, it is entered at the beginning of that month. Sometimes there have been sales for which the month is not known; in such cases I have entered these at the beginning of the year prefixed by N.d. (No date). Unfortunately there are other sales for which even the year is not known, or at most there is only the slightest indication. These have been grouped alphabetically in a separate section at the end of the *Register*.

If the name of the vendor of a sale is known but not the date, the index at the end of the work should be consulted and the page reference to the vendor's name will lead to the date and particulars of the sale.

In auctions held over a number of days but of which only the date of the first day of the sale is specified on the sale catalogue, the remaining days are indicated by ffdd (following days).

COLUMN 2 gives information relating to the source of the material offered for sale, including the name of the vendor (if known) and sometimes the place of residence. Broadly speaking these particulars are transcribed direct from the sale catalogue or, in its absence, from the authority (usually a bibliographical reference) shown in Column 5.

Occasionally the vendor's name is added in manuscript in the catalogue of an anonymous sale. When that is so the name is included in the *Register* in square brackets.

COLUMN 3 gives the contents of the sale. An unbracketed figure following a category, or categories, indicates the total number of natural history lots in a sale in which there are also irrelevant items. The absence of any figure or 'Irr' signifies as a rule a purely natural history sale. A figure in brackets indicates specifically the number of lots for the particular category or categories that it follows. In a general sale of more than one day the relevant day for the indicated lots is designated numerically, e.g. 21.xii (21st December).

COLUMN 4 gives the name of the auctioneer followed by the total number of lots in each sale and the number of pages and plates in the sale catalogue. A house sale is shown in italics in this column and its location appears in Column 2. The locality of a provincial sale (other than that of a house sale) also appears in italics, and except for certain house sales the absence of any locality in this column signifies a London sale.

COLUMN 5. In this column are the references to sale catalogue locations. If no catalogue is on record, then the authority for the sale entry, e.g. Da Costa, (1812:514) is shown. A catalogue location in brackets shows that the catalogue there has been searched for but has not been found. A catalogue location printed in bold type shows that I have seen the catalogue that it is there.

I have assessed the number of buyers' names and extent of pricing entered in marked catalogues as follows: p = prices, n = names (of buyers), m = many, s = some, f = few. Abbreviated bibliographical references appearing in this column are to be found in full at the end of one or other of the contributory articles.

1710 – 1759	SOURCE	CONTENTS	AUCTIONEER & SALE CAT.	REFERENCE
1710 March 22 ffdd	BERNARD (Charles), Serjeant Surgeon to her Majesty	Lib	Ballard 3540 lots 8 + 222pp.	**RCS** p.
1724 May 30 (postponed from 22 v 1724)	MARLOW (Jeremiah), retired goldsmith, late of Lombard Street	Min 172	Luffingham 305 lots 14pp.	**BML**
1728 November 11 ffdd	WOODWARD (John),[1] F.R.S., Fellow College of Physicians, Prof. of Physic in Gresham College†	Cor Hort. sicci Fish Rept Sh Ins Fos Cab Nh, 36	Bateman & Cooper 5092 + 61 bis lots 4 + 8 + 287pp.	**BML**, B p.
1750 N.d.	JONES (Thos. Wm.), Beaufort Buildings, Strand	Sh Fos	Langford, *house sale*	Da Costa (1812:514)
May 19-20	BOERHAAVE	Sh (118) Min (81)	Goldmann 207 lots 14pp.	(SR)
1751 May 24-27	RICHMOND (Duke of)†	Nh, 105	Heath 254 + 1 bis lots 14pp.	**BMM** s.p., SR
1753 N.d. [at least six days duration]	LISLE (Commodore)	Sh	Langford	Da Costa, *Elements of Conchology* (1776:204) Dance (1966: 230)
1755 March 11-15	MEAD (Dr. Richard) [M.D., F.R.S., b. 1673 d. 1754]	Nh, 72	Langford 368 lots 15pp.	**BMM**
1756 June 16-17	FURZER (Daniel), "late of New-Inn"†	Nh, 63	Langford 187 + 2 bis lots 10pp.	**BMP**
1757 June 4 ffdd	SADLER (Thomas) [d.ca. 1754 *teste* Da Costa (1812:205)]	Sh, 133	Langford 289 lots 12pp.	(CEB)
1759 April 27	STEELE (Edward), painter	Fos Min Sh Nh, 13	Paterson 126 + 1 bis lots 4pp.	**BML**

[1] This was mainly an auction of books the sale of which alone occupied 29 days. Also included in the sale were Woodward's cabinets of foreign and 'additional' English fossils and numerous irrelevant items (cf. Eyles, *J. Soc. Bib. nat. Hist.*, 1971, **5**(6), 416, 426)

1759 – 1767	SOURCE	CONTENTS	AUCTIONEER & SALE CAT.	REFERENCE
May 18-24	POND (Arthur), Great Queen Street, Lincoln's Inn Fields†	Sh Fos Cor (630) Nh (9) Lib (23) Cab (6)	Langford 669 + 2 bis lots 23pp.	CI s.p.f.n.
October 25-26	BARRINGTON (Mrs)	Nh	Paterson	(BSMB)
1760 January 20-21	AMES (Joseph), secretary to Society of Antiquaries† [b. 1689 d. 1759]	21 ii: Sh Fos Min Cor	Langford 192 lots 10pp.	**BMM** p.s.n., SR, VA
November 25-27	BAKER	Sh Ins Nh, 318	Paterson 320 lots 10pp.	(SR)
1762 April 7-8	LAWRENCE (Andrew), Pallmall, apothecary†	Sh Fos Cor Min, 28	Langford 175 lots 12pp.	**BMM**
November 27-28	CALDECOT, surgeon, antiquarian, Peterborough†	27 xi: Sh Min Fos Cor	Bristow 336 lots 16pp.	**BMM**
1765 April 29	HAMILTON (Matthew)	Sh Min Nh	Paterson 338 + 5 bis lots 10pp.	(SR p.n. add.); (G)
May 9	NORTHCOTE (Thomas), housekeeper of Hick's Hall†	Nh, 55	Hobbs 282 lots 12pp.	(SR)
1766 April n.d.	MAYNE (David)	Scottish fos	Paterson	Da Costa (1812:514)
April 9-12	(a) BLANCKLEY, apothecary, Crutched Friars† (b) an 'Antiquarian'	9 vi: Min Sh Fos Nh, c.50	Bristow 464 lots 20pp.	**BMM**; F
May 15-16	STUKELEY (William), M.D., F.R.S., F.A.S., Fellow of College of Physicians.	15 v: Fos Min, 21	Paterson 236 + 2 lots 14pp.	**BMM**; SR
June 3-4	JEANS (John)	Min[1]	Paterson 146 lots 7pp.	F
June 5-6	POCOCKE (Dr), Lord Bishop of Meath†	Min Sh Cor[2]	Langford 207 lots 12pp.	RCS
1767 January 19-20	ANON	Sh Nh	Paterson 178 lots 7pp.	(SR)

[1] Collected by Jeans in north of Scotland, many lots showed the locality or county in which the specimens were found.

[2] Richard Pococke (1704-1765). His min and fos were 'partly collected in his Scottish travels' (*Dict. Nat. Biog.*, **46**: 14)

1767 – 1774	SOURCE	CONTENTS	AUCTIONEER & SALE CAT.	REFERENCE
June 2-3	FLEMING (George), painter, Wakefield†	2 vi: Fos Min, 17	Paterson 197 lots 8pp.	BMM; CEB
December 14-18	FRANCKCOMBE (Wm), barrister, late of the Middle Temple†[1]	Fos Min Sh	Paterson 568 lots 20pp.	RCS s.p.
1768 March 5-7	ROUBY (John, Jam.) M.D.†	Fos Cor Sh	Paterson 251 lots 12pp.	RCS
June 14-15	THEOBALD (James), F.R.S., F.A.S.†	Fos Sh Ins Min Nh, c. 30	Langford 152 + 2 bis lots 10pp.	BMM, SR
1769 May 15-20, 22-23	FLUDYER (Sir Thomas), Lee, Kent†	17 v: Min Cor Fos Sh Nh, 70. 18 v: Fos Min Nh, 101 19 v: Ins Min Fos Sh Nh, 73 20 v: Nh, 24	Browning, *house sale* 899 lots 48pp.	BMM
1770 April 4-5	STURGES (John), architect	Nh, 29	Bastin 185 lots 8pp.	(BSMB), (CEB)
June 7-8	ANON.	Nh	Bastin 215 lots 10pp.	(SR)
December 17-18	HUGHES (Robert), 'Consul General for the Dutch, and Agent for the English, at Grand Cairo, Egypt'	Sh Fos Cor Min, 95	Bastin 162 + 2 bis lots 9pp.	SR, BMNHL copy
1771 June 3	'An Eminent Collector'†	Fos Sh Nh.	Christie 111 lots 8pp.	C p.n.
June 6	GRACE (Thomas), Devonshire Square†	English & For birds	Christie 69 lots 4pp.	C
1773 May 24	DYER (Samuel), F.R.S.	Sh Fos Seaweeds	Christie 92 lots 8pp.	C p.n., BMM
1774 April 27	BURROUGH (Rev. Henry), Prebendary of Peterborough†	Fos Sh Nh, 5	Gerard 95 lots 7pp.	BMM

[1] William Franckcombe (born Bristol, August 6th 1734, died September 3rd 1767) was 'an accurate and learned fossilogist, chiefly in petrifacta and had a numerous and well-chosen collection of fossils' 'Mr Ingham Foster bought his diary or catalogue of observations on fossils, a MS.' (E.M. Da Costa, *Gent Mag,* (1) **1812**, 206).

1775 – 1778	SOURCE	CONTENTS	AUCTIONEER & SALE CAT.	REFERENCE
1775 January 23-24	ANON	Fos Min Ins Nh, ca.30	Warre 255 lots 11pp.	BMM s.p.
March 13-18, 20-23	BAKER (Henry), F.R.S., F.A.S.†	Fos Min Sh For. plants & seeds, Cor Ins Nh.	Paterson 1189 lots 2 + 50pp.	BMM; SR; BSMB
[November] N.d.	JACKSON, dealer	Sh Irr.	Langford 480 lots	Henry Seymer *litt.* to Richard Pulteney in Pulteney Correspondence in Linnean Soc. (cf. Dance, 1966:108)
November 7-8	OWEN (Arthur), Llangwelach, Glamorganshire†	7 xi: Ins in amber, 1	Gerard 173 lots 8pp.	BMM p; SR
November 24	ANON	Sh	Christie 110 lots 7pp.	C p.n.
December 6-9	GRANT (Baron)	9 xii: Sh, 6	Christie 388 lots 22pp.	C p.n.
1776 February 9-10	COLEBROOKE (Josiah)†	Nh, 178	Langford 180 lots 11pp.	(SR)
November 4-12	'A Gentleman'	10 xi: Sh, ca.20	Christie 792 lots 41pp.	C p.n.
1777 April 9-11	'A Lady going abroad'	Sh Min Cor Seaweeds, Nh, 191	Willoughby 309 lots 16pp.	NMW m.p.m.n.
June 18-20	GOSLING (Rev. William), minor canon, Canterbury	Nh, 171	Langford 252 lots 11pp.	(SR)
December 18-19	'A Gentleman'†	18 xii: Min Cor Sh Butts, 5	Gerard 168 + 3 bis lots 8pp	BMM p.
1778 March 20-23	'A Lady'†	20 iii: Fos Min Nh, 8	Christie 377 lots 19pp.	C p.n.
March 28 – April 10	BOSC DE LA CALMETTE (Colonel), Maestricht	Fos Min Sh Cor Rept Fish Nh, 1843	Gerard 2108 lots 2 + 97pp.	BMM
September 4-5, 7	'A Physician and F.R.S.'	Min Sh Cor Fos	Paterson 510 lots 22pp.	RCS

1778 – 1782	SOURCE	CONTENTS	AUCTIONEER & SALE CAT.	REFERENCE
October 12-17, 19-22	FALCONAR (Magnus)†	Min Sh Fish Rept Ins Birds Bats Frogs Nh Irr	Paterson 1029 lots 4 + 40pp.	**BMNHZ** p.m.n.; **GLAS**
1779 February 25	'A Person of Distinction gone abroad'	Min Sh, 35	Christie 117 lots 6pp.	**C** p.n. add.; **BMM**
April 5-10, 12-17, 19-24, 26-30, May 1, 7-8, 10-15	HUMPHREY (George)	Sh Fos Cor Rept Ins Min Nh.	Paterson 4310 + 9 bis lots 2 + 168pp.	**H** p; **BMNHZ**; **BMNHZ** p.n. copy (original in Oslo Public Library, Oslo)
December 2-4	ANON	2 xii: Sh Seaweeds Irr	Christie 276 lots 10pp.	**C** p.n.
1780 March 30	'A Gentleman'†	Sh Min Ins Nh, 47	Paterson 171 lots 7pp.	**BMM**
April 6	MORRIS (Richard), 'late of the Tower'†	Sh Cor Fos Min Nh, 44	Burton 86 lots 7pp.	**BMM**
May 10-11	ROPER (John)	Sh Cor Nh, 123	Paterson 201 lots	(SR)
May 25	[BEAUCLERK (Hon. Topham) F.R.S.][1]	Min, 7	Paterson	**F** inc.
1781 June 14-15	'An Officer belonging to His Majesty's ship The Discovery, lately arrived'	Sh Ins Birds Nh Irr.	Hutchins 248 lots 15pp.	**NMW** copy
August 1	PENROSE (Rev. Thomas)†	Sh Cor Min Nh Fos, 556	Paterson 700 lots 26pp.	(SR)
1782 January 28 – February 13	MILLAN (John), Part I (Natural History)	Nh	Hutchins 1680 lots 73pp.	(SR)
February 6-7	DEVISME (William)†	Sh Min Chinese lep, 79	Christie 261 lots 13pp.	**C** p.n.
April 25-27, 29	[YEATS (Thomas Pattinson)]	Cayenne birds (457) Quad Rept	Hutchins 486 + 2 bis lots 22pp.	**BML**
May 6	JAMES (Colonel)	Sh Min, 63	Christie 76 lots 6pp.	**C** p.n.

[1] Only a fragment of the catalogue remains and the title and names of auctioneer and vendor are entered in pencil.

1782–1786	SOURCE	CONTENTS	AUCTIONEER & SALE CAT.	REFERENCE
June 10-17	'A Gentleman'	Nh, 2	Christie 902 lots 41pp.	C p.n.
1783 March 10 ffdd	FOSTER (Ingham)[1]	Fos	Barford	Da Costa (1812:515)
May 12	YEATS (Thomas Pattinson), F.R.S.	Birds Ins Sh Nh	Hutchins	Da Costa (1812:515)
May 15 – June ffdd	FOSTER (Ingham)[2]	Sh Cor Cab Nh	Barford	Da Costa (1812:515)
1784 May 20	FOSTER (Ingham)	Sh Fos Lib	Egerton	Da Costa (1812: 515)
November 2-5	[LALANDE (Dr)]	Sh Min, 343	Chapman 450 lots 22pp.	(SR)
December 8-13	'A Person of Distinction'	Sh Fos, 5	Christie 570 lots 26pp.	C p.n.
1785 N.d.	MATTHEWS (Rev. Thomas)	Sh Nh	Greenwood	Da Costa (1812:516)
February 4-5	[LYNN]	5 ii: Rept Nh Irr	Hutchins 188 lots 12pp.	**BMNHZ** p.
March n.d.	SPEED, druggist, Cannon Street	Sh	Hutchins	Da Costa (1812:515)
May 11-12	ANON[3]	Sh Cor Min	Greenwood	**CLD** inc.
December 21-23	'A Collection in the Country'	Sh Min Ins Birds Irr	Christie 361 lots 19pp	C p.n.
1786 N.d.	POND (Arthur) [but collected by Jacob NEILSON (*teste* Da Costa *loc. cit.*)]	Sh Fos	Langford	Da Costa (1812:513)
April 24-29, May 1-6, 8-13, 15-20, 22-27, 29-31, June 1-3, 6-8, July 3	PORTLAND (Duchess Dowager of), Privy Garden, Whitehall†	Lep Col Hem Sh Cor Min Ool Birds Cab Mam Herb Nh Irr	Skinner, *house sale* 4156 + 104 lots 8 + 194 + front-ispiece + 6pp.	**RCS**; **BMNHZ** p; **H**; **BMP**; **FM** p.n.
August 16-18	NEILSON (Jacob)†[4]	Sh Nh Fos	Hutchins	(SR)

[1] Catalogued by E.M.da Costa, this sale lasted 10 days and realised £317. 1s. (Da Costa, *loc.cit.*)

[2] Catalogued by E.M.da Costa, this sale lasted 28 days and realised £646. (Da Costa, *loc.cit.*)

[3] The whole of the catalogue for the second day is missing, but the title shows the sale to have been entirely nat. hist., of 2 days duration and totalling 131 lots for the first day.

[4] Da Costa (1812:513) says George Humphrey catalogued it, and 'Isaac Swainson bought many Sheppey crabs'.

1787 – 1789	SOURCE	CONTENTS	AUCTIONEER & SALE CAT.	REFERENCE
1787 March 30	'A Nobleman'	Sh Ins Min Cor Birds	Christie 137 lots 10pp.	C p.n. add.
December 10-15, 17-22, 24	'An Eminent Professor of Anatomy'	21 iii: Recent & Fossil Animal skeletons, 32	Hutchins 1328 lots 4 + 139pp.	BML
1788 March 12-14	WHITE (John), Part III (Natural History)	12 iii: Sh Cor Fos Ins Min Birds, 125	Gerard 350 lots 20pp.	BMM p.n.; SR
April 26-28	BOND (Benjamin), Clapham, Surrey†	Sh Min S.American ins Exotic & American birds, 40	Christie 234 lots 15pp.	C p.n.
April 30 – May 2	[KING (Dr)]	2 v: Min Sh, 33	Gerard 414 lots 14pp.	BMM p.n.
May 23-26	ANON	26 v: Sh Min Fos Exotic ins, 110	Christie 283 lots 12pp.	C p.n.; BML
December 12	ANON	Nh, 3	Greenwood 128 lots 8pp.	DIP; **BMNHL** copy
1789 March 21, 23	FITZROY (Lady)	23 iii: Sh Fos Nh, 43	Christie 231 lots 12pp.	C p.n.
May 4-8	ARDESOIF (John), Tottenham, Middlesex†	6 v: Birds Nh, 15	Christie 524 lots 28pp	C p.n.
May 5-6	[DOUGLAS (Lady)] †	Sh Cor Min Fos Ins Nh	Hutchins 243 lots 20pp.	H p.s.n.
May 7	[DOUGLAS (Lady)] †	Min Sh Lib	[Hutchins] 117 lots	H inc.[1]
May 15	[ROPER]	Sh Min Fos Cor	Hutchins 122 + 1 bis 8pp.	H m.p.s.n.
June 8-11	(a) JACOB (Edward), F.A.S., Feversham (b) LIGHTFOOT (Rev.), Chaplain to the Dowager Duchess of Portland†[2]	8-10 vi: Fos Sh Cor Min Bot Birds Ool Ins Fish Rept Lichens Nh, 364	Gerard 473 + 2 bis lots 26pp.	SR; **BMNHL** copy

[1] Cat. lacks title. Page 1 is inscribed: 'This was a collection made up for sale after Lady Douglas's [catalogue] was printed, sold as by auction'.

[2] Jacob was the author of *Sheppey Fossils* (1777); and Lightfoot, of *Flora Scotica* (1775).

1790 – 1794	SOURCE	CONTENTS	AUCTIONEER & SALE CAT.	REFERENCE
1790 January 13-14	ANON	13 i: Herb (1), Min (17)	Gerard 202 lots 8pp.	BMM
March 8-9	MINARD (Mrs)†; et al.	Sh Cor Min Ins Nh, 268	King & Chapman 298 + 4 bis lots 12pp.	SR; **BMNHL** copy
March 10-11	RAWLE (William), Castle Court, Strand†	10 iii: Min Sh Nh, 8	Hutchins 195 lots 12pp.	BMP; CEP
1791 May 9-11	(a) MANNERS (Lord James)† (b) 'A Gentleman'†	Min Fos Sh, 221	Christie 365 lots 19pp.	C p.n.; VA
May 26-28	WALKER†	Min Fos Sh Cor Nh, 126	Christie 369 lots 16pp.	C p.n. add; VA
June 10	ELLIS (John), F.R.S., F.A.S.[1]	Sh Fos Min Rept Seaweeds Cor Nh Zoophytes	Hutchins 107 lots 12pp.	BMP p.f.n.
September 1-3	RENNIE (W.), engraver †	Sh Fos Min, 26	Greenwood 375 + 1 bis lots 16pp.	CLD
1792 March 12-13	BURTON (Philip)	13 iii: Sh Min Fos Nh, 15	Christie 241 lots 20pp.	C p.n.
March 16-17	PARSONS (James), M.D., F.R.S., F.S.A.†	16 iii: Min Fos Sh Nh, 39	Gerard 196 lots 8pp.	SR; **BMNHL** copy
May 14-15	CONSTABLE (William),† his museum chiefly formed by Marmaduke TUNSTALL, Yorkshire†	Min Sh Cor Ins Fos, 132	Christie 212 lots 14pp.	C p.n.
May 22-24	BARKER (Sir Robert)†	Sh Min, 75	Greenwood 392 + 11 bis lots 24pp.	CLD
1793 June 3	BUTE (John, Earl of), Luton House, Bedfordshire†	Min Fos Cor Nh Irr	Skinner & Dyke	Turner, 1967, *Ann. Science*, 23:242 plt XII
1794 January 11, 13-14	'A Museum'	11 i: Sh Min Nh, 15	Christie 269 lots 14pp.	C p.n.; BML

---

[1] Ellis was the author of *History of Zoophytes* (1786). The sale included drawings of madrapores, sponges and other zoophytes by Roberts, Miller, Taylor, G.D.Ehret and others. Lot 30 consisted of a 'wainscot box containing a very large collection of fuci or sea-weeds' the whole of which was arranged and labelled by Ellis. Lot 97 (purchased by Humphreys) consisted of 'seven large glazed frames... in which are arranged a very fine and extensive collection of corals and corallines, sponges, etc., from which the figures and descriptions in the *History of Zoophytes* was taken; most of the specimens, which are chosen ones, are labelled either by Mr. Ellis or Dr. Solander'.

1794 – 1797	SOURCE	CONTENTS	AUCTIONEER & SALE CAT.	REFERENCE
May 2	JACKSON (John), F.S.A., Clement's Lane†	Min Fos Nh, 34	King 118 lots 4pp.	BMP
May 8-10, 12-17, 19	STUART (John, 3rd Earl of Bute)[1]	Lib Hortus siccus	Sotheby 1257 + 28 bis lots 4 + 64pp.	LINN; BML p
May 23-24	ANON	Sh Min Nh, 40	Gerard 208 + 2 bis lots 12pp.	BMP
May 30	'A Gentleman (leaving off collecting)'	Min, 60	King 135 lots 7pp.	BMM
June 2	'A Gentleman left off collecting'	Ins, 4	Hutchins 87 lots 7pp.	BMM
August 15	'A Gentleman,' Ratcliffe	Sh Min Ins Crustacea Cab, 70	King 136 lots 8pp.	BMM s.p.
1795 April 30	LEAKE (William), M.D.†	Sh Nh, 9	Christie 102 lots 12pp.	C p.n.
May 12-13	SOUTHGATE (Rev. Richard), A.B., Rector of Warsop, Nottinghamshire†	Fos (34) Sh (184)	Sotheby 226 + 2 bis add. lots 12pp.	BML p.n.
July 3	'An Eminent Physician'	Birds	Christie 189 lots 8pp.	C p.n. add.
1796 April 29 – May 2	[RUMBOLD (Sir Thomas)]	1 v: Sh Nh, 74	Christie 409 lots 18pp.	C p.n.; CLD; FK
August 5 (2nd day)	ANON	Min Cor Fos Sh, 84	Wells & Fisher (?) 106 (108-213 lots) 4 (pp. 9-12)	SR; **BMNHL** copy
October 27-29, 31, November 1-2	LEATHES, apothecary, George Street, Hanover Square†	2 xi: Lib Seaweeds Bot. drawings, 7	Sotheby 1564 lots 2 + 50pp.	BML p.n.
1797 June 9	ANON	Birds, 50	Phillips 150 lots 10pp.	P p.n.

[1] This sale contained three Hortus siccus including one (lot 1255) 'lately the property of Gronovius'.

1798 – 1800	SOURCE	CONTENTS	AUCTIONEER & SALE CAT.	REFERENCE
1798 June 13-15	ANON†, Part II (Natural History)	Sh Cor Min	King 436 lots 29pp.	BMNHZ
June 22	WOSCOTT (Rev. J.)	Sh Nh, 6	Christie 96 lots 8pp.	C p.n.
July 24-25	ANON	Sh Min Nh	King 300 lots 22pp.	BMNHZ
1799 January 2-4	'A Dignified Emigrant'	Sh Cor Ins	King 378 lots 24pp.	BMNHZ
February 15-16	BRAAM (A. E. van), chief director Dutch East India Co., Canton	16 ii: Min Nh, 60	Christie 126 lots 10pp.	C p.n.
April 22-25	ANON	Sh Min Cor Fos	King 506 lots 31pp.	BMNHZ: NMW f.p.
April 29	'RACKSTROW'S MUSEUM'	Sh Cor Fos Min Rept Birds Quad	King 159 lots 8pp.	BMNHZ
April 30	'RACKSTROW'S MUSEUM'	Birds Quad	King 171 + 1 lots 10pp.	BMNHZ
May 2-4, 6	BAYLY (G.), Piccadilly†, Part II	Sh Fos Birds Min Ins Nh	King 452 lots 20pp.	BMNHZ
November 21	MOHR	Min	King 50 lots 4pp.	(SR s.p.)
1800 January 14	'Purchased at Rome'	Nh, 3	Christie 40 lots 4pp.	C p.n.
February [18]	WALKER, surgeon†	Min Fos Sh Nh, 90	Leigh & Sotheby 121 + 5 bis lots 9pp.	BML p.n.
March 10-14	MORTON (Charles), M.D., F.R.S., Principal Librarian, British Museum†	14 iii: Min Fos, 25	Sotheby 1149 + 25 lots 2 + 35pp.	BML p.n.
April 30 – May 1	TASSIE (James)†	Min Irr	Phillips 300 + 2 lots 23pp.	P p.n.
May 20	PEARSON (Mr. & Mrs.), Highgate	Bird drawings, 11	Phillips 77 lots 8pp.	P p.n.

1800 – 1802	SOURCE	CONTENTS	AUCTIONEER & SALE CAT.	REFERENCE
July 8-9	'A Foreigner going abroad'	Min Ores Fos	Christie 298 lots 14pp.	C p.n.
August 20-21	SLEATH, 'retiring from business'	2 viii: Nh, 12	Christie 202 lots 15pp.	C p.n.
September 3-6, 9	CHICHESTER (Marquis of Donegal)	Sh Cor Birds Fos Nh	King 783 lots 42pp.	BMNHZ
October 30-November 1	(a) 'A Lady' (b) ANON	(a) Sh Cor Fos Min Ins Nh, 352 (b) Birds, 31	King 383 lots 21pp.	BMNHZ
1801 February 9-11	[HUNTER (Mr.)]	9 ii: Min, 27	Sotheby 519 + 3 bis lots 2 + 17pp.	BML p.n.
March 10-14, 16	[CHICHESTER (Marquis of Donegal] †	Min (440); Sh Cor Fos (110)	King 640 lots 26pp.	BMNHZ; BML; BMM
April 25	BUTE (John, Earl of)	Bot drawings Sh drawings[1]	King 73 lots 6pp.	BMP
May 25-30, June 1-6, 8-13, 15-20, 22	CALONNE (Charles Alexandre de), Leicester Square	Sh Min Fos Cor Birds Fish Mam HH Ins Nh Irr	King, *house sale* 3037 lots 170pp.	BML p.
June 8-9	ANON	Min Sh	King 210 lots 12pp.	BMNHZ
November 27-28	'A Collector'	Sh Min Seaweeds Nh, 25	King 199 lots 8pp.	BMM p.
December 18	'A Gentleman'	Sh Min Cor Ins, 113	King 118 lots 7pp.	H p.
December 22-23	ANON	Sh Cor Min	King 260 lots 16pp	BMNHZ
1802 April 21	ANON	Min	King 130 lots 10pp.	B s.p.
May 24 – June 1	PICKARD (Leonard), York, Part II	29 v: Sh Fos Min, 93	King 1326 lots 2 + 56pp.	BMP; BML; SR

Drawings by Van Rozin, Edwards, De Heer, Withoos, Ehret, Merian, Jewel, Van Huysum, Taylor, Ditche, King, Marcous and Moninck.

1803 – 1804	SOURCE	CONTENTS	AUCTIONEER & SALE CAT.	REFERENCE
1803 January 10-15, 17-18	SWINBURNE (Henry)	10 i: Min, 14	Sotheby 1586 + 4 bis lots 2 + 58pp.	BML p.n.
January 26-28	ANON	Sh Min Fos Nh, 386	King 420 + 6 bis lots 28pp.	B
April 18	ANON	Sh Min Fos Fish Nh, 25	King 86 + 2 bis lots 8pp.	BMM
June 17-18, 20	DAVALL (Edmund), F.L.S., Orbe, Switzerland	Lib (including much bot.)	Leigh & Sotheby 568 + 2 bis lots 2 + 19pp.	LINN; BML
June 24-25, 27-28	BRITISH MUSEUM, duplicates	Cor Min Cab, 416	Sotheby 439 lots 2 + 22pp.	BML p.n. add.
July 9	CALONNE (de)	Min Sh Cor, 5	Christie 59 + 1 bis lots 6pp.	FK; BMNHL copy
August 22-27	PARLBY (Rev. S.) Stoke near Nayland, Suffolk†	Min Sh Cor	King 795 lots 64pp.	BMNHZ
October 25	ANON	Sh Min Ores, 86	King 140 lots 8pp.	B
[November] after 15 ('in a short time')	SHELDON (Thomas)	Sh Fos Cor Cab	King	Advert on p.7 King Cat. of an irr. sale of 15.I 1803 in BML (ref. SC 813(40)
1804 January 7, 9-14	'The property of a foreign prince consigned from Germany'	Min Sh	King (pp.1-8, 25-31)	BMNHZ inc
April 11-12	SUMNER (John), Brompton†	11 iv: Cor Min Fos Lep Birds Quad, 87	Christie 188 lots 16pp.	C s.p.
April 16-21	ANON	Min	King 752 + 12 bis lots 50pp.	B
April 30 – May 5	ANON	Sh Min Ins Nh	King 769 lots 2 + 59pp.	BMNHZ
May 15-16	'A Gentleman'	16 v: Sh Fos Fish Birds Min Nh, 71	King 401 + 1 bis lots 16pp.	BRAD
May 29-30	LIPTRAP (John), bankrupt, Mile End	Min (13) Sh (30) Nh (15)	Sotheby 213 lots 15pp.	BML

1804 – 1806	SOURCE	CONTENTS	AUCTIONEER & SALE CAT.	REFERENCE
June 16-18	ANON	Sh Min	King 413 + 8 bis lots 22pp.	**BRAD**
December 12-14	[COHEN]	14 xii: Sh Min, 4	Leigh & Sotheby 517 lots 21pp.	**BML**
1805 May 23-25	DRURY (Dru), goldsmith, Strand†	Ins Lib	King & Lochée 343 + 4 bis lots 16pp.	**H** p.n. add.; **BMNHL** p.n.add. copy; **BMNHE** p.n.
July 16	(a) GREEN, Westminster† (b) - (c) ANON[1]	(a) Col Dipt Hym Br lep, 32 (b) Sh Min Cab, 72 (c) Lib, 4 (d) Fos Irr	King & Lochée 144 lots 8pp.	**H** p.f.b. **BMNHL** p.f.b. copy
1806 January 15-16	ANON†	16 i: Sh Min Cor Nh, 25	King 212 lots 11pp.	**B**
January 29	ANON	Min Sh, 23	King & Lochée 139 + 2 bis lots 8pp.	**B**
March 18-21	'A Gentleman'†	Sh Min Ins Fos Birds Nh	King & Lochée 515 lots 44pp.	**BRAD**
May 5-10, 12-13	LEVERIAN MUSEUM, Part I	Mam Min Sh Birds Nh Irr	King & Lochée Lots 1 - 960 41pp (1-2 + 1-39)	**RCS** p.n.; **LINN**, **BMNHZ**; **BRAD**; **CAMB**; M.m.p. m.n.
May 14-17, 19-22	LEVERIAN MUSEUM, Part II	Rept Min Mam Sh Nh Irr Birds	King & Lochée Lots 961 - 1920 44pp. (40-83)	**RCS** p.n.; **LINN**; **BMNHZ**; **BRAD**; **CAMB**; M.m.p. m.n.
May 23-24, 26-31	LEVERIAN MUSEUM, Part III	Sh Min Fos Ins Nh Irr	King & Lochée Lots 1921-2880 41pp (1-2 + 84-122)	**RCS** p.n.; **LINN**; **BMNHZ**; **BRAD**; **CAMB**; M.m.p. m.n.
June 2-7, 11	LEVERIAN MUSEUM, Part IV	Min Birds Sh Ins Nh Irr	King & Lochée Lots 2881-3840 62pp (1-2 + 123-182)	**RCS** p.n.; **BMNHZ**; **LINN**; **BRAD**; **CAMB**; M.m.p.m.n; **BMNHM**
June 12-14, 16-20	LEVERIAN MUSEUM, Part V	Birds Fos Sh Min Nh Ins	King & Lochée Lots 3841-4800 42pp. (1-2 + 183-222)	**RCS** p.n.; **LINN**; **BMNHZ**; **BRAD**; **CAMB**; M.m.p. m.n.

MS note on fly-leaf of cat. reads: 'Only the Insects were Green's and these had been previously in the possession of Martyn, formerly of Marlborough Street' 'The fossils belonged to a gentleman in Ireland'.

1806 – 1808	SOURCE	CONTENTS	AUCTIONEER & SALE CAT.	REFERENCE
June 21, 23-28, 30 - July 5, 7-9	LEVERIAN MUSEUM, Part VI	Birds Fish Mam Ool Nh Irr	King & Lochée Lots 4801-6840 76pp. (1-2 + 223-296)	RCS p.n.; LINN BMNHZ; BRAD CAMB; M.m.p. m.n.
July 11-12, 14	LEVERIAN MUSEUM, Part VII	Birds Min Sh Fos Nh Ins Irr	King & Lochée 354 lots 19pp.	RCS p.n.; LINN BMNHZ; BRAD CAMB; M.m.p. m.n.
July 14-18	LEVERIAN MUSEUM, Appendix[1]	Sh Cor Min Fos Nh Irr	King & Lochée 684 lots 35pp.	RCS p.n.; LINN BMNHZ; CAMB BRAD + 62pp. printed priced cat.
1807 February 19-20	ANON	Birds Min Sh Fish Rept Mam Nh, 243	King & Lochée 256 + 5 bis lots 12pp.	BRAD s.p. (E. Donovan's purchases)
March 14	REHE (S.), mechanist to the Admiralty	Birds, 9	Christie 15 lots 4pp.	C s.p.
March 30 ffdd	[DRAKE]†	Min Fos Birds Sh Lep Nh, 1136	Tregoning, *Penryn, Cornwall* 2829 + 48 bis + 381 + 2 bis lots 128 + 16pp.	BRAD; BMNHL copy
April 13-15	ROBSON (Col.), F.A.S., Lieut. Gov. of St. Helena†	Sh Min Fos Nh, 160	Abbott 379 + 6 bis lots 25pp.	BMM
April 13-18, 20-21	VON RUPRECHT, 'Chemical Professor at Schemnitz, Hungary'	Min	King & Lochée 1082 lots 44pp.	B
June 2-3, 5-6	INNOCENT, 'retiring from business'	2, 5 vi: Min Fos Sh, 201	Christie 544 lots 44pp.	C s.p.n.
June 13, 15-20	PIERSON (Rev. Archdeacon), Rector of Easingwould, Yorkshire	20 vi: Sh Min Fos, 49	Jeffery 1443 lots 56 pp.	BML
June 22-23	GRAVE (R.)†	22 vi: Sh Min, 43	Richardson 278 lots 14pp.	BMP; BUP
November 2-3	WOODFORD (Rev. Archdeacon)†	Exotic & other plants	Mant, *Winchester* 263 lots 15pp.	BRAD
1808 N.d.	GIESECKE (Sir Charles Lewis)	Min	Brodie, *Leith*	Allan, *Charles Giesecke* MS[2]

[1] Among the many purchasers at the Leverian sale were Bullock, Babington, Clift, Donovan, Haworth, Macleay, Milne, Parkinson, Pennant, Swainson and Tankerville.

[2] The 'contents of 9 or 10 casks' auctioned in the Spring of 1808 at The Shore, Leith (Allan, *Charles Giesecke* MS., copy to me from Miss J.M.Sweet *per* E.C.Pelham-Clinton – C.-H.) Also, cf. J.F.Johnstrup, *Gieseckes Mineralogiske Rejse i Gronland*, 1878; T.Allan, *Ann.Phil.*, 1813, **1**: 99-110, 217, **2**: 389; idem, *Trans. Roy. Soc. Edin.*, 1812, **6**: 345-351; T.Thomson, *Trans. Roy. Soc. Edin.*; 1812, **6**: 371-386; W.V. & K.R.Farrar, *Ann.Sci.*, 1968, **24**: 115-120; J.M.Sweet, *Ann.Sci*, 1972, **28**: 298.

1808 – 1809	SOURCE	CONTENTS	AUCTIONEER & SALE CAT.	REFERENCE
February 29 – March 2	BREE (Martin)	2 iii: Min Sh, 13	Leigh & Sotheby 613 lots 20pp.	BML p.n.
March 16-19, 21-23	HEULAND (Henry)	Min	King & Lochée 700 lots 57pp.	BMNHM p.
April 9	SULLIVAN (Sir Richard)†	Min, 1	Christie 19 lots 4pp.	C s.n.; VA; RKDH
May 2-7, 9-12	FORSTER (Jacob)[1], Gerard Street, Soho,† Part I	Min Sh Echini Cor Fos Nh	King & Lochée Lots 1 - 1206 98pp. (1-98)	BMNHM m.p.
May 19-21, 23-28, 30	FORSTER (Jacob), Gerard Street, Soho,† Part II	Min Sh Cor Fos Nh	King & Lochée Lots 1207 - 2490 119pp. (99-218)	BMNHM m.p.; BMNHZ
June 2-4, 6-11, 13	FORSTER (Jacob), Gerard Street, Soho,† Part III	Min Sh Echini Cor Fos Nh	King & Lochée Lots 2491 - 3790 104pp. (219-322)	BMNHM m.p.; BMNHZ
June 8-15	'An Eminent Jeweller and Lapidary'	Min Sh Irr	Christie 1200 lots 62pp.	C s.p.n.
June 14	LAMBERT (Brice)†	Ins	King & Lochée 110 lots 7pp.	H; BMNHL copy
June 16-18, 20-25, 27	FORSTER (Jacob), Gerard Street, Soho†, Part IV	Min Sh Echini Cor Fos Nh	King & Lochée Lots 3791 - 5090 84pp. (323-406)	BMNHM m.p.; BMNHZ
June 28-30 – July 2, 4	FORSTER (Jacob), Gerard Street, Soho†, Part V	Min Fos Cor Cab Sh Nh	King & Lochée Lots 5091 - 5860 52pp. (407-458)	BMNHM m.p.
July 19-20	FIELD†	20 vii: Min Sh Ins	King & Lochée 291 + 2 bis lots 12pp.	BMP
July 29	[LASKEY (Captain)]	Min Sh Fos Nh, 107	Squibb 120 lots 8pp.	BMNHZ; H
1809				
January 12-14, 16-18	STEVENS (James), Camerton	Lib	Leigh & Sotheby 1587 lots 56pp.	BML
March 27-30	EWER (Samuel), Hackney	30 iii: Min, 111	Sotheby 803 lots 2 + 32pp.	BML p.n.

Cf. P.J.P.Whitehead, Some further notes on Jacob Forster (1739-1806), mineral collector and dealer, *Min. Mag.*, 1973, **39**: 361-363. His brother Ingham Forster (or Foster) had sales in 1783 and 1784 (*q.v.*).

1809 – 1811	SOURCE	CONTENTS	AUCTIONEER & SALE CAT.	REFERENCE
April 18	MADDISON (John), 'of the Foreign Dept. in the Post Office'†	Sh Min, 69	King & Lochée 137 lots 7pp.	**BMM** p.n.; **B**
May 11-13, 15-20, 22-23	WOODFORD (Emperor John Alexander)	23 v: Drawings of birds plants and ins., 4[1]	Sotheby 1773 lots 2 + 78pp.	**BML** p.n.
June 8-10, 13-15	WILSON (W.), the Minories	Sh Min Fos Cor	Murrell 590 lots 2 + 49pp.	**BMP**
June 28	BARFORD (Henry)†	Sh, 1	Willock 132 lots 8pp.	CEP; **BMNHL** copy
1810 February 16-20, 22-23	DAVIS (T.C.)†, ANON.	23 ii: Min, 4	Leigh & Sotheby 1638 + 4 bis lots 55pp.	**BML** p.n.
May 15-16	MENISH (H.), M.D., Bishop's Hall, near Chelmsford, Essex.	Sh Min Fos Rept Cor Nh, 260	Kelham, *house sale* 303 + 1 bis lots 20pp.	**B**
May 17-19, 21-23	WEBBER (William)†[2]	Sh, 577	King & Lochée 836 lots 35pp.	**BMM** p.n.; **B** inc.; MDB p.n.
May 30–June 1	'An Eminent Collector'	30-31 v: Fos Sh Min Cor Nh, 13	Dodd 434 + 13 bis lots 20pp.	**VA**
June 7	ANON	Sh Min Rept Nh, 76	Christie 159 lots 12pp.	**BMM** p.n. add.
June 19	BELLAMY (Clement)†	Min Fos, 14	Christie 179 lots 10pp.	C p.m.n.; **VA**; RKDH
August 31 – September 1	ANON	Sh Min Nh, 16	Dodd 265 + 1 bis lots 12pp.	CEP; **BMNHL** copy
1811 January 3-4	WELCH (Mrs. Ann), Aylesbury†	3 i: Min, 91	Sotheby 162 lots 2 + 9pp.	**BML** p.n.
January 16-17	ANON	Sh Cor Ins Min Birds Nh, 222	King & Lochée 282 + 2 bis lots 14pp.	**BMM**; **B**
February 28 ffdd	BERESFORD (John Claudius)	Lib Irr	Jones [*Dublin*] 1121 + 3 lots 2 + 74pp.	**VA**

[1] Lot 1773 'a magnificent and unique collection of ornithology consisting of eighteen hundred drawings and prints of birds by Lewin, Sydney Edwards and R.R.Reinagle . . . in 12 volumes large folio . . .' Bought by Dent for £378, these included drawings from Bankes Mus., Capt. Cooke and Lever Mus.

[2] Lot 556, a very large Wentletrap from Amboyna bought by Bullock for £27. For further information on this sale vide Dance, *Shell Collecting* (1966:231).

1811 – 1813	SOURCE	CONTENTS	AUCTIONEER & SALE CAT.	REFERENCE
March 26-31, April 2-3	LETTSOM (John Coakley), M.D., F.R.S., Camberwell.	Lib	Leigh & Sotheby 1347 lots 52pp.	**BML**
May 1-4	WHITE (Joseph), Islington†	Sh Min Cor, 362	King & Lochée 492 + 5 bis lots 37pp.	**BRAD**
May 8-9	'A Gentleman'	Min Cor Rept Sh For birds Br ins, 184	Christie 234 lots 12pp.	**C** p.n.
May 9-10	ZOFFANY (John), R.A., artist	10 v: Sh Nh, 17	Robins 206 + 1 bis lots 12pp.	**CLD**; VA inc.
May 31 – June 1	'A Gentleman'	1 vi: Sh, 49	Robins 277 + 1 bis lots 19pp.	**VA**
1812 February 8	BRUNTON (James), 'Colonel Adjutant General of the Forces at Madras'	English birds, 69	Christie 171 lots 10pp.	**C** p.n.
February 10-12	FENTON (Ibbetson), Walworth, lately deceased, Member of the Entomological Society†	Lib	Leigh & Sotheby 604 lots 22pp.	**BML**
April 29	SMITH (Mathew), Col., F.A.S.†	Cor Min Nh, 4	Christie 155 lots 12pp.	**C** p.n., **BMM**
May 4-7, 11-14	HEULAND (Henry)	Min Fos	Christie 1127 lots 80pp.	**C** p.n.; **BMNHM**
May 25	'A Man of Science'	Fos Min Nh, 145	Christie 195 lots 14pp.	**C** p.n.; BML
June 2-3	CAVENDISH (Frederick), Market Street, Herts†	3 vi: Seaweeds from near Harwich Min, 3	Leigh & Sotheby 646 lots 18pp.	**BML**
1813 January 5	ANON	Sh Min Birds Ins, 138	King & Lochée 159 lots 8pp.	**BRAD**
February 19-20	ANON	Sh Cor Min, 224	King & Lochée 255 + 1 bis lots 14pp.	**B**
February 22-24	GOLDING (Henry), Wallingford, Berks.†	Sh Min Fos Ins Birds Cor Nh Nh drawings, 250	King & Lochée 336 + 2 bis lots 15pp.	**H**
March 23-24	ANON	Sh Cor Fish Echini Birds Nh	King & Lochée 308 lots 12pp.	**BRAD**

1813–1814	SOURCE	CONTENTS	AUCTIONEER & SALE CAT.	REFERENCE
May 4	(a) – (b) ANON	(a) Birds (b) Dup. birds	King & Lochée 242 + 1 bis lots 14pp.	BRAD
June 16	'A Gentleman'†	Min Fos, 1	Denew 113 lots 7pp.	VA; CLD
June 25-26	'A Foreign Nobleman'	25 vi: Sh Min, 17	Christie 162 + 38 lots 10pp.	C p.n.; RKDH; VA; FK
July 2	LINNEAN GALLERY formed by Dr. THORNTON	Bot drawings Portraits of botanists Bot lib.	Christie 90 lots 7pp.	C p.n.
November 15-16	'A Gentleman'	Sh Min Ins Fos Birds, 277	King & Lochée 314 lots 13pp.	BRAD
December 20-21	WILLETT (Ralph)†	Bot drawings Nh drawings Irr	Sotheby Lots 2716-2906 19pp. (101-119)	B
1814 March 7-8	ANON	Min Sh, 156	Ballantyne, *Edinburgh* 227 + 1 bis lots 8pp.	(SR)
May 5	Postponed till 9 v 1814			
May 6-7	ANON	7 v: Fish Birds Ins Min Nh, 173	King & Lochée 332 + 3 bis lots 14pp.	BMM
May 9 (postponed from 5 v. 1814)	TALBOT (Sir Charles)†	Min Fos	Sotheby 95 + 1 add lots 7pp.	BML p.n.
May 11-13	WILLETT (Ralph), Merly†	Drawings[1]	Philipe 318 lots 2 + 15pp.	BMP; VA; SR
May 18-21, 23	WOODD (John), Old Burlington Street†	18-21 v: Min Fos Sh, 626	Squibb 769 lots 42pp.	BMM
June 14	[LATHAM (Dr. John), Romsey][2]	Ins, 218	King & Lochée 218 + 11 bis lots 8pp.	H; BMNHL copy
July 9	[MacLEAY (Alexander)]	Dup ins.	King & Lochée 122 lots 8pp.	H

[1] Drawings by G.D.Ehret of sh, bot and a few ins; and drawings by Lang of Nuremburg of birds.

[2] There are two copies of the cat. in the Hope Dept., one of which bears E.Donovan's MS. annotations: 'I lotted and gave the names of the Insects in this catalogue, Dr. Latham having bestowed very little attention on the Science of Entomology, and requesting as a favour that I should name them'. There is also a MS. note on the title: 'Written by me in conformity with Dr. Latham's wishes E.D.'

1814 – 1816	SOURCE	CONTENTS	AUCTIONEER & SALE CAT.	REFERENCE
November 11	PHILIPS (John Leigh)†, Part II	Ins	Winstanley & Taylor, *Manchester* 88 lots 14pp. (76-89)	BMP; VA
November 25	'A Man of Distinguished Taste'	Min, 29	Christie 164 lots 8 + 2 pp.	C p.n. add.
December 24	ANON	Sh Min Ins Nh, 38	King & Lochée 169 lots 8pp.	B
1815] February 15-18, 20-22	ANON	21-22 ii: Drawings, 44	Richardson 805 lots 2 + 34pp.	VA; SR
1815 February 16-17	(a) – (b) ANON	(a) 16 ii: Min Sh Birds, 163 (b) 17 ii: Sh Min Fos, 3	King & Lochée 265 lots 14pp.	BMM
March 20	[BENTINCK (Count)] †	Sh Min, 23	Sotheby 144 + 1 bis lots 2 + 8pp.	BMM p.n.; BML p.n.
April 10-15, 17	BOURBON (Duke de), 27, Orchard Street, Portman Square†	13 iv: Sh Cor Fos, 96 Birds 14 iv: Min, 119	George, *house sale* 791 + 2 bis lots 67pp.	BRAD
April 24-29	HEULAND (Henry)	Min	King & Lochée 820 lots 44pp.	BMNHM
June 9-10	ANON	Fos Min Sh Birds, 232	King & Lochée 348 lots 15pp.	BMM; B
November 2-3	ANON	2 xi: Sh, 6	Christie 193 lots 20pp.	C p.m.n.
1816 January 15-18	[COOMBE (Dr.)]	Sh	Sotheby 442 lots 2 + 24pp.	BML p.n.
January 26-29	LLWYDIUS (=LLOYD)	Nh [+ Irr], 214	Broster 2356 lots 128pp.	(SR)
February 28	ANON	Ins Nh, 147	King 153 + 1 bis lots 8pp.	BRAD
March 13-15	ANON	Min Cab, 262	King 355 + 1 bis lots 18pp.	BML
March 25-30	HEULAND (Henry), 25, King Street, St. James's	Min (part I)	King 748 + 9 bis lots 46pp.	BMNHM

1816–1817	SOURCE	CONTENTS	AUCTIONEER & SALE CAT.	REFERENCE
April 1-5	HEULAND (Henry)	Min (part II)	King 600 lots 34pp.	**BMNHM; BML**
April 22-27	HEULAND (Henry)	Min (part III & last)	King 816 + 2 bis lots 46pp.	**BMNHM**
May 2-4	LETTSOM (John Coakley)†	4 v: Fos Sh Min Nh, 93	Leigh & Sotheby 359 lots 24pp.	**BML** p.n.
June 6-8	BRITISH MUSEUM	Dup. min (part I)	Christie 363 lots 45pp.	**C** p.m.n.; **BMNHM** inc.
June 25-26	BUCKINGHAMSHIRE (Albinia, Countess of)†	26 vi: Sh, 13	Christie 364 lots 34pp.	**C** p.n.
July 11-17	JENNINGS (Henry Constantine)	Sh Cor Min Birds Fish Echini Amphibia Ins, 865	Phillips 937 lots 53pp.	**P**
September 9-13	JENNINGS (Henry Constantine)	Nh, 120	Phillips 544 lots 44pp.	**(P)**
November 19-21	WILKIN (Simon)[1], Cossey near Norwich	20 xi: Lib, 135. 21 xi: Ins Fos Cab Sh Mol Entomological app. Illustrations, 210	Basham & Harman, *house sale* 607 + 10 bis lots 29 + 6pp.	**NOR; BMNHL** copy; **BML**
1817 February 20-21	ANON	Min Sh Fos, 293	King 304 + 1 bis lots 12pp.	**BML**
April 9	ANON	Birds Sh Ins Nh, 188	King 241 lots 8pp.	**H**
April 24	ANON	Sh Min Rept Fish Fos Ins Nh, 165	King & Lochée 176 + 1 bis lots 8pp.	**BRAD**
May 27-28	FRANCILLON (J.), 24, Norfolk Street, Strand†	Fos Min Sh[2] Cor Rept Lib Br ool Exotic birds App. Nh	King 184 lots 20pp.	**H; BMNHL** copy
July 25-26	FRANCILLON (J.), 24, Norfolk Street, Strand†	Dup. ins	King 253 lots 12pp.	**H** p.n. (of 1st day only); **BMNHL** p.n. (of 1st day only) copy
July 25-29	'A Lady'†	Nh, 73	Phillips 720 lots 41pp.	**(P)**

[1] In *History of the B.M. (N.H.) Collections,* under Vigors, it is stated that Wilkin's British coleoptera were bought by Vigors and that some portions went to the B.M. (N.H.) in a selection of the Vigors coll. in 1859. It contained type specimens of a few species (*teste* K.G.V.Smith *in litt.* 26 XI 1974).

[2] Including lot 55: 'an extensive collection of British Shells formed by Mr. Cranch'.

1817–1819	SOURCE	CONTENTS	AUCTIONEER & SALE CAT.	REFERENCE
August 20-22, 25, 27-29	ASTLEY (Francis Dukinfield), Dukinfield Lodge, near Ashton-under-Line (*sic*), Cheshire	25 viii: Birds, 31	Winstanley & Crole, *house sale* 856 lots 66pp.	VA
1818 March 10-14	BONELLI (Angelo)	11-12 iii: Sh, 60	Phillips 750 lots 51pp.	P
March 16-18	'A Gentleman gone abroad'	16 iii: Birds, 4	Winstanley 347 lots 27pp.	VA
April 30 – May 2, 4-8	DONOVAN (E.) 'London Museum and Institute of Natural History', Part I	Min Fos Cor Birds[1] Seaweeds Bot Rept HH Vermes Echini Zoophytes Fish Crustacea Lep Ins Mam	King 878 lots 2 + 56pp.	CAMB; BMNHL copy
May 2	ANON	Nh	Phillips 188 lots 14pp.	(P)
June 10-11	'A Gentleman'	10 vi: Sh, 30	Christie 301 lots 24pp.	C p.m.n.
June 11-13, 15-20	FRANCILLON (John)†	For ins Arachnidae	King 1328 lots 2 + 74pp.	BMNHE; (LNSW)
June 22-26	JENNINGS (Henry Constantine)†	Nh, 150	Phillips 722 lots 31pp.	(P)
November 10-11	ANON†	10 xi: Sh, 3	Christie 264 lots 20pp.	C p.m.n.
1819 April 11-13	TOWNSEND (Rev.)	Min Fos	King 325 lots 16pp.	J
April 29-30, May 4-7	BULLOCK'S LONDON MUSEUM OF NATURAL HISTORY, Part I[2]	4-6 v: Birds, 356 7 v: Quad, 57	Bullock 664 + 3 bis lots 4 + 40pp.	BMNHZ p.n.; VA; BML; CAMB p.n.
May 11-14, 18-19	BULLOCK'S LONDON MUSEUM OF NATURAL HISTORY, Part II	Sh Fos Birds Quad	Bullock 849 + 1 bis lots 42pp.	BML; BMNHZ p.n.; VA; CAMB p.n.
May 14-15	HEWETT (W.N.W.)	Nh 'collected in India', 19	Phillips 273 lots 19pp.	(P)

Lot 649, 'Great Auk, *Alca impennis* . . . this is a British specimen, and the original bird described by Pennant, Dr. Latham, and all later authors'.

Cf. J.M.Sweet, *Ann.Sci.* 1970, **26**: 23-32 re the University of Edinburgh's part in this sale.

1819–1821	SOURCE	CONTENTS	AUCTIONEER & SALE CAT.	REFERENCE
May 20-21, 25-28	BULLOCK'S LONDON MUSEUM OF NATURAL HISTORY, Part III	Birds Sh Rept Fish Cor Ins	Bullock 741 lots 36pp.	BMNHZ p.n.; VA; CAMB p.n.
June 1-4	BULLOCK'S LONDON MUSEUM OF NATURAL HISTORY, Part IV	Birds Quad Ins Fos Nh	Bullock 548 + 13 bis lots 27pp.	BMNHZ p.n. CAMB p.n.
June 8-11	BULLOCK'S LONDON MUSEUM OF NATURAL HISTORY, Part V	8-10 vi: Birds Nh	Bullock 546 + 4 bis lots 25pp.	BMNHZ p.n. add. CAMB p.n.
June 15-19	BRODERIP (William), Bristol	Sh Min Cor Echini Asteria Nh	King 752 + 11 bis lots 40pp.	LINN m.p.f.n.
June 21	ANON	Min Sh Ins in amber, 25	Bullock 99 lots 19pp.	BML
September 8-10	MARSHAM (Thomas)	Col Orth Hem Neur Lep Hym Dipt Arachnidae Acera	King 466 lots 2 + 31pp.	BML; CH copy (LNSW)
December 7-11	ANON	Sh Min Fos Birds Ool Ins	Bullock 555 + 2 bis lots 31pp.	CAMB
1820 January 20 ffdd	PLAYFAIR (John), F.R.S.L.	Min Irr	Ballantyne, *Edinburgh* 1421 lots 4 + 106pp.	VAE; NLS
May 1-6	HEULAND (Henry)	Min	King 775 + 3 bis lots 39pp.	BML p.n.; BMNHM p.n. copy
May 15	BIRCH (Colonel)[1]	Fos	Bullock 102 lots 1 + 12pp. + 1 plate	BMNHP; GEOL CAMB p.m.n.
1821 February 27-28, March 1-2, 6-7	DICK (P.), Sloane Street	27 ii; 2, 7 iii: Fos Min Sh Animal skulls Rept Birds Ins Quad Nh, 237	Bullock 830 lots 56pp.	BMM
March 26-31, April 2-5	ROYAL MUSEUM, 28, Leadenhall Street, London	Birds Rept Quad Marine crustacea Nh, 1138	Dubois 1299 lots 76pp.	J
May 21-26, 28-30	ANGUS (Mrs.)[2]	Sh Cor	Dubois 1410 + 3 bis lots 79pp.	LINN p.n.; NMW
August 1-11	WRIGHT (Dr), F.L.S., M.G.S., Lichfield	Sh Min Rept Fish Nh, 929	Harris, *Lichfield* 1546 + 5 bis lots 43pp.	RCS; SR

[1] The fossils were from the blue lias at Lyme and Charmouth, Dorset. It was subsequently understood that all the valuable fossils in this sale were obtained by the indefatiguable labours of Miss Mary Anning. The plate is of lot 40 'A crustaceous insect approaching to the Palaemon of Olivier'. The sale included the Ichthyosaurus described by Sir Everard Home and figured by Clift in Phil.Trans. for 1819 – it realised £152. 5s. (cf. Mantell, *Lond.Geol. Jnl.*, **1846**: 13).

[2] Stated in the cat. as having been 'principally from Mr. De Calonne's Museum'. Nearly all the lots were of shells.

1822 – 1825	SOURCE	CONTENTS	AUCTIONEER & SALE CAT.	REFERENCE
1822 March 11-12	PLYMPTON (Richard)	11 iii: Nh, 21	Phillips 260 lots 8 + 8pp.	(P)
May 20-25	BLIGH (Mrs.)†[1]	Sh	Dubois 1024 lots 4 + 58pp. + coloured frontispiece + appendix 20pp. + coloured frontispiece.	**LINN** p.; **NMW** p.; **BMNHZ**
1823 February 28	ANON	Sh Min, 14	Jones 189 lots 8pp.	**BMM**
June 10-13	(a) SWAINSON (William), F.R.S., Part I (b) – (c) ANON	(a) For birds For ins. (b) Indian ins. (c) Birds & Ins. of New Holland	Dubois 972 lots 2 + 37pp.	**BMNHZ**
September 17 add	LYNCH PIOZZI (Mrs. Hester)†	22 ix: Min Fos, 26	Broster, *Manchester* 1205 lots 44 + 24pp.	NYK; **BMNHL** copy
September 9-20	'A Private Gentleman'	Sh Nh Irr	Jones 417 lots 15pp.	VA
1824 February 24	ANON	Nh	Christie 191 lots 12pp.	(C)
April 29-30	ANON†	Nh, 45	Christie 264 lots 14 + 3pp.	(C)
April 30 – May 1, 4-5	(a) HEULAND (Henry), incl. Marquis de DREE (b) ANON	(a) Min, 601 (b) Min Fos, 3	Dubois 602 + 2 bis lots 35pp.	**BMNHM** p. copy; BML p.
May 14	MILNE (George), F.L.S.	Dup. lep. Dup col.	Thomas 176 lots 14pp.	H f.p.; **BMNHL** f.p. copy
May 28-29, 31	D'ONIS (The Chevalier), late Ambassador from Spain	Min, 442	Phillips 438 lots 32pp.	W p.n. add.
1825 N.d.	BULLOCK'S MEXICAN EXHIBITION	Fish Birds Irr	Thomas	Allingham (1924:26)
May n.d.	STEPHENS [James Francis]	Dup ins.	Stevens	(LNSW)

[1] Mrs. Bligh (née Elizabeth Betham) was the wife of the well-known Admiral Bligh of the *Bounty*. The catalogue was by William Swainson. For further information on this sale cf. *Jnl. Conch.,* **20**: 154 (1935)

1825 – 1828	SOURCE	CONTENTS	AUCTIONEER & SALE CAT.	REFERENCE
May 5	ANON	Sh Min Fos, 1	Southgate 121 lots 8pp.	BML p.n.
1826 March 7-10	HAWORTH (Richard)	Lib Irr	Sotheby 1035 lots 33pp.	JR p.n.; BML p.n.
April 10-15	HEULAND (Henry), incl. Marquis de DREE	Min	Thomas 1150 + 22 bis lots 59pp.	BMNHM
April 26-29	BARNES (William), Redlad Hall near Bristol†	28 iv: Birds Ins, 11	Sotheby 365 + 6 bis lots 25pp.	BML p.n.
June 28	TUKE (Dr.)	Min Sh Nh, 279	Jones, *Dublin* 916 lots 36pp.	(RIA)
1827 N.d.	STUTCHBURY (Samuel), A.L.S.			*Zool.Jnl.*, **5**:99 but possibly not an auction sale
February 19-20	ANON	Bot drawings, 12	Southgate 283 lots 25pp.	BML p.n. add
April n.d.	PARKINSON (James)[1]	Fos		Br. Mus. (N.H.) *History of the Collections*, **1**: 315
April 20 ffdd	CRICHTON (Sir Alexander), F.R.S., M.D.[2]	Min	Sowerby 2600 + 121 bis lots 128pp.	BMNHM m.p.
November 20-21	[BULLOCK]	Min Sh Fos Mam Nh, 262	Thomas 310 + 3 bis lots 18pp.	B
1828 February 29 – March 1	VILLET (M.) 'Zoological Museum, Egyptian Hall, Piccadilly'	Birds Ins Mam Fish Sh Nh	Thomas 345 lots 24pp.	BMNHZ p.n.
May n.d.	ANON[3]	Fos	[Thomas]	Allingham (1924:25)
July 14-19, 22-26, 29	BROOKES (Joshua), F.R.S., F.L.S., F.Z.S., 'Anatomical and Zoological Museum', Part I	18-19, 23, 29 vii: Mam Bats HH Fos, 329	Robins 859 + 2 bis lots 76pp (1-76)	RCS; BMNHZ

[1] Cf. Sowerby, *Min.Conch.*, **6**: 112; also, Clark and Hughes, *Life and Letters of Adam Sidgwick* (1890), 280. Mantell (*Lond. Geol. Jnl.*, **1846**: 14) called it a 'matchless collection'.

[2] The BMNIIM cat. formerly belonged to F.N.A.Fleishmann. The sale continued for 16 days.

[3] Of fossils 'a remarkable collection' sold in May 1828 (Allingham, *loc.cit.*) with extracts from the cat. (idem, 44-46) including 'some Leverian material'.

1828 – 1830	SOURCE	CONTENTS	AUCTIONEER & SALE CAT.	REFERENCE
July 30 – August 2, 5	BROOKES (Joshua), F.R.S., F.L.S., F.Z.S., 'Anatomical and Zoological Museum', Part II	30-31 vii, 1 viii: Birds, 487	Robins 864 + 2 bis lots 47pp. (77-123)	**RCS**
September 20-24	RIVERS (Lord), Eltham Lodge, Kent†	21 ix: Birds Mam, 18	Christie 553 + 4 bis lots 42pp.	**C** p.n. add.
1829				
May 4-9	HEULAND (Henry)	Min	Thomas 1020 lots 60pp.	**BMNHM** p.n.
May 26-27	SIMS (John), M.D., F.R.S., Dorking	Herb Bot lib, 491	Thomas 725 + 1 bis lots 28pp.	**BMNHB**
June 19	CALEY (George), Curator, Botanic Garden, St. Vincent†	Herb Bot lib Birds, 143	Christie 148 lots 18pp.	**BMNHB**
October 6-10, 12	HEAVISIDE (J.) 'Heaviside's Museum', George St., Hanover Sq., Part II (& last)	Ins Birds Quad Fish Nh Irr	Wheatley & Adlard, *house sale* 2397 lots 2 + 267pp.	**BMNHZ** m.p. m.n.
November 30 – December 5	(a) SIRR (Major Henry), Dublin. (b) ANON	(a) Sh Min Cor, 43 (b) Irr	Wright, *Edinburgh* 732 lots 28pp.	**NLS**; **BMNHL** copy
1830				
February 18-20, 22-25	HENDERSON (John)†	20 ii: Animal skulls Fos Min, 3	Sotheby 1148 + 1 bis lots 55pp.	**BMM** p.n.; **VA**; **CLD**
March 1-5, 8-12, 15-19, 22-26, 29-31	BROOKES (Joshua), F.R.S., F.L.S., F.Z.S., 'Anatomical and Zoological Museum', Part III & appendix	11-12, 15-22, 29 iii: Birds Fish Quad Rept Fos Mollusca Entozoa Ins Nh Irr	Wheatley & Adlard 2712 + 2 bis lots 132 + app. 4pp.	**RCS** m.p.m.n.; **LINN**; **CAMMUS**
April 19-22	BOWLES (William), Fitzharris House, Abingdon, Berks.	Min	Thomas 800 lots 49pp.	**GEOL**
April 5-24	[RANDALL?]; et al.	5-7 iv: Lib, 620; 2 iv: Min Herb Butts Cab, 47	Tait, *Edinburgh* 2836 + 6 bis lots 96pp.	**NLS**; **BMNHL** copy
May 10-12	HEULAND (Henry)	Min	Thomas 611 + 1 bis lots 37pp.	**GEOL**; **BMNHM** s.p.
May 31 – June 2	HEULAND (Henry)	Min	Thomas 510 + 3 bis lots 31pp.	**BMNHM** f.p.
June 11-15	STUART (General Charles), Bengale†	12 vi: Nh Irr	Christie 553 + 1 bis lots 59pp.	**C** m.p. add.
June 15-16	WESTON (Rev. Stephen), B.D., F.R.S., F.S.A.†; ANON	16 vi: Min Fos Birds, 14	Sotheby 227 + 5 bis lots 16pp.	**BMM** p.n.; **BML**

1831–1833	SOURCE	CONTENTS	AUCTIONEER & SALE CAT.	REFERENCE
1831 May 2-5	HEULAND (Henry)	Min	Thomas & Stevens 714 + 1 bis lots 50pp.	BMNHM s.p. add
May 25-26	ANON	26 v: Min 4	Sotheby 605 lots 20pp.	BML p.
August 3	ANON	Fos Sh Nh, 61	Southgate 64 lots (1811-1874) 4pp.	RCS
November 15-18	TRATTLE (Marmaduke)†	18 xi: Min Sh, 125	Hoggart 465 lots 21pp.	VA p.n.
1832 May 14-17	HEULAND (Henry)	Min	Thomas & Stevens 715 lots 50pp.	BMNHM p.
May 18	ANON[1]	Ins Fos Min Bot MSS on fos, 55	Hodgson 72 + 5 bis lots [4] pp.	BMNHP inc.
1833 March 14 ffdd	CLERK (John), Eldin†	27 iii: Fos Nh, 2	Winstanley, *Edinburgh*	GLA
April 15-19, 22 ffdd	HARRINGTON (Henry), 'Harrington Gallery and Museum', Part III	15-19 iv: Sh, 610	Gernon, *Dublin* 2110 + 4 bis lots 2 + 21 + 51pp.	VA
April 29 – May 4	HEULAND (Henry)	Min	Thomas & Stevens 1200 + 6 bis lots 73pp.	BMNHM m.p. f.n. add.
May 27 – June 1, 3-4	SOWERBY (G.B.)	Sh (the whole of his private coll. and stock)	Thomas & Stevens 1665 lots 4 + 91pp.	BMNHZ
June 9-10	SOWERBY (G.B.)	Fos Min Lib		Advert on p.91 of Thomas & Stevens Sale Cat. of 27 v - 4 vi. 1833
July 22 ffdd	HARRINGTON (H.)	Min	Lindley 1220 lots 2 + 40pp.	GEOL; BMP

---

[1] A MS. note on p.1 of the cat. reads: 'Contained part of the Lever Collection & ?John Woodward's MS. & Lwyd's'. Lot 43, 'A Descriptive Catalogue (Cat. Raisonee) of all the Petrefactions or extraneous Fossils in the Leverian Museum... by Emmanuel Mendez da Costa... No part of this Catalogue has ever been published... This invaluable MS. is contained in 9 Fasiculi, folio'.

1833 – 1835	SOURCE	CONTENTS	AUCTIONEER & SALE CAT.	REFERENCE
October 14	ANON	Sh Min Birds, 125	Thomas & Stevens 261 + 1 bis lots 10pp.	BMM
1834 January 27	ANON	Sh Min Fos, 15	Southgate 203 lots 9pp.	BML p.n.
February 27 dd	PAYNE (John Howard); ANON	4 iii: Min, 8	Sotheby 2132 lots 72pp.	BML p.n.
March 10-13	DELAFONS (J.P.)	Sh Ins Rept Min Fos Birds Fish HH Mam Crustacea	Thomas & Stevens 885 lots 38pp.	BMNHZ s.p.s.n.; CAMMUS
March 11-13	HAWORTH (Adrian Hardy)†	Ent lib Bot lib	Sotheby 737 + 1 bis lots 29pp.	H m,p.s.n. (Westwood's copy); BMNHB
April 14-16	HEULAND (Henry), incl. Dowager Countess of AYLESFORD†	Min	Thomas & Stevens 610 + 3 bis lots 40pp.	BMNHM m.p.m.n.
May 5-9	HEULAND (Henry), incl. Dowager Countess of AYLESFORD†	Min Cab	Stevens 840 + 2 bis lots 56pp.	BMNHM m.p.m.n.; GEOL
June 23-28, 30– July 4	HAWORTH (Adrian Hardy)†[1]	Br & For ins (incl. Col Lep Neur Dipt Hem Hym Orth Strepsiptera Dermaptera Aptera) Cab	Stevens 2306 lots 1 + 105pp.	GS; J; H
August 12-16	ANON	Sh Min Fos Nh, 127	Southgate 1220 + 2 bis lots 35pp.	BML p.n.
November 14-15	ANON	15 xi: Birds (60), Nh (12)	Southgate 367 + 1 bis lots 18pp.	BML p.n.
1835 January 8	SOWERBY (G.B.), 50 Gt. Russell St., Bloomsbury	Sh Fos Min Lib Cab	Sowerby, *house sale* 117 lots 16pp.	GEOL
May 11-15	HEULAND (Henry), incl. Dowager Countess of AYLESFORD†	Min	Stevens 1002 + 7 bis lots 59pp.	BMNHM m.p.
November 11-14, 16-21	FORSYTH (William), F.A.S., F.S.A., gardener to his Majesty; FORSYTH (William), Nottingham Place.	Bot lib Nh lib	Sotheby 2597 lots 91pp.	BMNHB

At a meeting of the South London Entomological and Natural History Society on 24 iv 1890, a priced example of the sale cat. was exhibited (cf. *Proc. S. Lond. ent. nat. Hist. Soc.*, **1890-91**: 27).

1835–1837	SOURCE	CONTENTS	AUCTIONEER & SALE CAT.	REFERENCE
November 25	ANON	Nh drawings[1]	Sotheby 144 + 1 bis lots 9pp.	BML
1836 February 1-6, 8-13, 15-19	SIVRIGHT (Thomas), Meggetland and Southouse†	15 ii: Min Sh	Tait, *Edinburgh* 215 + 2 bis lots 155pp. + 5 plates	W p; NLS
February 2 ffdd	MOTHERWELL	15 ii: Mam, 3	Stuart, [*Edinburgh*]	GLA
February 17	[FINNELL (Captain Benjamin)][2]	Fos (remains of mastodon mammoth of Chio etc.)	Stevens 200 lots 11pp.	**BMNHP** p. (of lots bought by BM trustees)
March 3-4	LE GRAND (George Wren)†	3 iii: Min Nh, 8	Sotheby 357 + 3 bis lots 21pp.	**BMM** p.n., VA **BML**
May 2-6	HEULAND (Henry), incl. Dowager Countess of AYLESFORD†	Min	Stevens 1000 lots 58pp.	**BMNHM** f.p.f.n add.
June 22-24	LEATHES (Rev. G. R.), Shropham Villa, near Larlingford, Norfolk†	Sh Fos Min Lichens Br Iep For ins.	Deck, *house sale* 480 lots 28pp.	**BMNHZ**
July 18-20	LEATHES (Rev. G. R.), Shropham Villa, near Larlingford, Norfolk†	Sh Ins Fos Nh	Stevens 610 + 10 bis lots 26pp.	**CAMMUS**; **SPD** copy
July 25-26	SURREY MUSEUM, Marlboro' Place, Walworth	[Nh Irr]	Stevens, *house sale*	Advert on p.26 of Stevens' cat. of 18-20 vii 1836
July 25-27	BURTON (James, jun.)	26-27 vii: Crustacea Rept Zoophytes Nh, 69	Sotheby 420 lots 36pp. + 3 plates	**BML** p.n.; VA
1837 April 18-21	SLATER (J.)†	2 iv: Min Irr	Christie 625 lots 24pp.	**C** p.n. add.
April 28	HEULAND (Henry)	Fos Min, 201	Stevens 200 + 1 bis lots 10pp.	**BMNHP**
May 18	TANKERVILLE (Charles Bennet, 4th Earl of)†	Min	Stevens 221 + 1 bis lots 16pp.	**GEOL**
May 24, 26	COBBE (Colonel), political agent, Moorshedebad†	Indian birds Min Nh, 135	Christie 156 + 1 bis lots 16pp.	**C** p.n.; CEP

[1] Lot 123 consisted of 2500 drawings of butterflies and moths in 4 vols. folio – the originals of P.Cramer's plates for *Uitlandische Kapellen*.

[2] For some prices realised cf. Mantell, *Lond. Geol. Jnl.*, **1846**: 14.

1837 – 1839	SOURCE	CONTENTS	AUCTIONEER & SALE CAT.	REFERENCE
June 12-16	HEULAND (Henry), incl. Dowager Countess of AYLESFORD†	Min	Stevens 1002 + 7 bis lots 59pp.	BMNHM m.p.
July 4-17	BROUGHTON CASTLE, near Banbury, Oxfordshire	4 vii: Sh Nh, 22	Enoch & Redfern, *house sale* 1143 lots 63pp.	VA; EKW
July 28	(a) BENZA (Dr.) (b) ANON	(a) Birds Mam Sh Nh (all from India & Ceylon) (b) Birds & Mam of Sumatra	Sotheby 223 + 7 bis lots 13pp.	BML p.n.
December 11-12	(a) 'A Gentleman in the East Indies' (b) HEULAND (Henry)	(a) 'Choice Iron Ores' (b) Min Cab	Stevens 507 + 1 bis lots 35pp.	BMNHM s.p.
1838				
n.d.	GUILDING (L.)	Sh	Stevens	Sherborn (n.d.)
May 14-17	HEULAND (Henry), incl. Dowager Countess of AYLESFORD†	Min	Stevens 837 + 1 bis lots 48pp.	BMNHM f.p.
June 6-8	CAPE OF GOOD HOPE ASSOCIATION FOR EXPLORING CENTRAL AFRICA SOCIETY'S MUSEUM	Birds Rept Quad Nh, 494	Stevens 559 + 2 bis lots 23pp.	BMNHZ m.p. m.n. add; **J**; **CAMB** f.p.; (LNSW)
June after 8	'A Lady'	Exotic sh	Stevens	Advert on p.23 of Stevens' cat. of 6-8 vi. 1838.
June 14	WINDELER (D.)	Min	Stevens	Advert on p.23 of Stevens' cat. of 6-8 vi. 1838.
June 20	HEULAND (Henry)	Min Ores	Stevens	Advert on p.23 of Stevens' cat. of 6-8 vi. 1838.
August 2	ANON	African nh Irr	Sotheby 101 lots 7pp.	BML p.n.
November 8	ANON	Fos	Stevens 64 lots 4pp.	BMNHZ
1839				
n.d.	CHRISTIAN OF WALDECK & PYRMONT (Princess)	Nh	Stevens	Allingham (1924:34)
February 18-19	HEULAND (Henry)	Min Cab	Stevens 407 + 16 bis lots 34pp.	BMNHM p.n. add.
May 7	WALKER (Sir Patrick)†	Br ins For ins	Stevens 207 lots 16pp.	H; BMNHL copy
May 9-10	ANON	Sh Birds Nh Irr	Fletcher, *Manchester*	SPD (printed prospectus (1 page) only)

1839–1841	SOURCE	CONTENTS	AUCTIONEER & SALE CAT.	REFERENCE
May 13-17	HEULAND (Henry), 31, Red Lion Square, incl. Dowager Countess of AYLESFORD†	Min Cab	Stevens 1048 + 131 bis lots 78pp.	BMNHM m.p.
June 18-19	MYLNE (George), F. L. S.	Exotic ins. (including Col Lep Hem Orth Neur Hym Dipt) Br ins	Stevens 386 + 2 bis lots 23pp.	H; BMNHL copy
1840 N.d.	GUILDING (L.)	Sh	Stevens	Sherborn (n.d.)
March 30 – April 4	CHILDREN (John George)	Br and For ins (including Col Neur Dipt Hem Hym Orth Lep) Cab	Stevens 950 lots 54pp.	J s.p.; H s.p. s.n. ('H' = Hope purchases); BMNHL s.p.s.n. copy
April 6-8	CHILDREN (John George)	Ent lib	Sotheby 661 + 5 bis lots 32pp.	H p.; BMNHL p copy
May 4-6	HEULAND (Henry)	Min Cab	Stevens 630 + 30 bis lots 43pp.	BMNHM s.p.s. n. (BM purchase only)
May 15-16	(a) – (c) ANON	15 v: (a) Sh Fos Min, 224 (b) Lep Col Nh, 15 16 v: (c) Min, 175	Stevens 488 + 1 bis lots 24pp.	GEOL
June 8	GOODALL (Rev. Joseph), late Provost of Eton	Sh	Stevens	Advert on p.2 of Stevens' cat. of 15 v. 1840.
July 16-17	SWAINSON (William)	Ins Birds Rept Lib Cab	Stevens 407 lots 20pp.	J f.p.
July 30-31	SEALE (R. F.) ['SCALE' in error] St. Helena Museum.	Mam Fish Birds Ins Sh Rept	Foster 327 lots 20pp.	EKW p.n., BMNHL p.n. copy
November 12	JANVIER (Dr.)	Fish	Stevens	*Athenaeum* 1840: 881
1841 April 29-30	BARTLETT (Rev. T. O.), The Rectory, Swanage	Birds (88) Fos (89) Min (12) Nh (11)	Bonfield, *house sale* 261 + 4 bis lots 16pp.	GEOL
May 10-11	ANON	Min	Stevens 456 lots 32pp.	BMNHM f.p.
November 24	BAUER (Francis), botanical painter to George III†	Bot lib Bot drawings Fos Sh Microscopic preparations Irr	Christie 216 lots 8pp.	C m.p. add.
December 11	ANON, Clapham	Min Sh Fos	Sotheby 174 lots 12pp.	BML p.n. (of min & fos only)

1842–1844	SOURCE	CONTENTS	AUCTIONEER & SALE CAT.	REFERENCE
1842 March 31 – April 1	McENERY (Rev. John)†	Fos Min, 176	Walke, *Torquay* 220 + 1 bis lots 11pp.	BMNHP m.p.
April 18-20	LAMBERT (A. B.), F.R.S., F.S.A.†	Bot lib	Sotheby 689 + 1 bis lots 42pp.	BMNHB p.n.
April 21	LAMBERT (A. B.), F.R.S., F.S.A.†	Cor Fos Bot Irr	Sotheby 71 + 2 bis lots 8pp.	BMNHB
May 9-12	HEULAND (Henry)	Min	Stevens 821 + 8 bis lots 51pp.	BMNHM f.p.
June 27-29	LAMBERT (A. B.) F.R.S., F.S.A., 26, Lower Grosvenor St., London†[1]	Bot Herb	Sotheby, *house sale* 321 lots 16pp.	BMNHB p.b.add.
September 22-24, 26-29	LANGSTAFF (G.)	28 ix: Rept Fish Birds Quad Animal skeletons, 206	Stevens 2027 lots 92pp.	RCS p.n.; BMNHZ p.m.n. add.
October 24-29	BAKER (George), Northamptonshire historian	29 x: Min Fos Sh Ins, 71	Sotheby 1510 + 28 bis lots 84pp.	BML p.n.; VA
December 3	YOUNG (Matthew)†	Sh Min Fos Cab, 2	Sotheby 202 lots 13pp.	BML p.n.
1843 January 19	HUGHES (E.), Lincoln's Inn	Sh Min Fos	Stevens	*Athenaeum*, **1843**: 26
May 15-17	HEULAND (Henry)	Min	Stevens 600 + 1 bis lots 34pp.	BMNHM f.p.f.n. (BM purchases only)
May 25	KOCK (Albert), St. Louis, U.S.A.	Remains of N.American mastodontoid animals Fos	Stevens 104 + 1 bis lots 15pp.	BMNHP
November 23	KOCK (Albert), St. Louis, U.S.A.	Remains of N.American mastodontoid animals Fos	Stevens 85 lots 12pp.	BMNHP
1844 May 2-3	(a) HOOKER (Joseph) (b) ENTOMOLOGICAL CLUB	(a) For ins Br ins (b) For ins	Stevens	*Athenaeum*, **1844**: 369
May 13-14	HEULAND (Henry)	Min	Stevens 420 + 2 bis lots 37pp.	BMNHM f.p.f.n. (BM purchases only)

[1] For an account of the Herbaria at this sale cf. *Lond. Jnl., Bot.*, **1**: 394.

1844 – 1846	SOURCE	CONTENTS	AUCTIONEER & SALE CAT.	REFERENCE
May 30-31	(a) BAKEWELL (Robert) (b) ANON	(a) Min Fos Lib (b) Sh Ins Birds	Stevens	*Athenaeum*, 1844: 466
June 20	ANON	Min Fos Sh	Stevens	*Athenaeum*, 1844: 538
July 4	ANON	Fos Min Sh Rept Birds Mam	Stevens	*Athenaeum*, 1844: 585
July 16	(a) HEULAND (Henry) (b) TAMNAU (Dr. F., jun.) Berlin	(a) Min (b) Fos	Stevens 245 lots 15pp.	BMNHM
July 25	HAWKINS (Thos.)	Fos (incl. ichthyosauri & plesiosauri)	Stevens 118 lots 7pp.	RCS
1845 February 17-19	STUTCHBURY (H.R.), bankrupt bookseller and naturalist	Sh Min Fos Birds Ins (his stock of these groups)	Stevens	*Athenaeum*, 1845: 161
March 27	(a) ANON (b) SCHOMBURGK (Sir R.)	(a) Sh Min Fos (b) Birds Herb	Stevens	*Athenaeum*, 1845: 281
April 15-19, 21-23	JOHNSON (James), 12, Dowry Parade, Hotwells, Bristol†	Fos Min Sh	Austin, *house sale* 778 lots 32pp.	BMNHP; GEOL BMNHL f.p.
May 29	BUTTS (Thos.)†	Exotic sh	Stevens	*Athenaeum*, 1845: 506
June 5 ffdd	GRAHAM	Sh	Stevens	*Athenaeum*, 1845: 506
June 16	HEULAND	Min Fos	Stevens	*Athenaeum*, 1845: 553
June 18	'An Eminent Author on Natural History'	Lib	Sotheby 104 lots 10pp.	BML p.n.
December 2-3	SALE (Richard Cowlishaw)†	2 xii: Fos Sh Min Cab, 19	Sotheby 280 + 1 bis lots 15pp.	BML p.n.
December 11	HUBERT (Henry)†	Min Fos Sh	Stevens	*Athenaeum*, 1845: 1161
1846 February 16-17	HEULAND (Henry)	Min	Stevens 400 lots 27pp.	BMNHM f.p.
March 5	WARBURTON (Dr.)	Br ins.	Stevens	*Athenaeum*, 1846: 209
March 23	(a) ANSTICE (Robert), F.G.S., Bridgewater† (b) ANON	(a) Sh Fos Min Nh, 46 (b) Sh Min Fos, 37	Sotheby 181 + 1 bis lots 10pp.	BML p.n.
April 6-7	GRAHAM (Robert), F.R.S., Regius Professor of Botany, Edinburgh University†	Bot lib (409) Herb (152)	Tait & Nisbet, *Edinburgh* 531 + 30 bis lots 24pp.	BMNHB m.p.s.r

1846 – 1847	SOURCE	CONTENTS	AUCTIONEER & SALE CAT.	REFERENCE
April 23	ANON	Sh Min Birds Nh	Stevens	*Athenaeum*, **1846**: 386
May 7-8	ANON	Sh Birds Ins Herb Nh Corallines Animal skulls	Stevens	*Athenaeum*, **1846**: 441
May 18-22	HEULAND (Henry)	Min Fos	Stevens 1002 lots 56pp.	BMNHM p.n. add
May 28-29	ANON	S.African animals	Stevens	*Athenaeum*, **1846**: 489
June 4-5	SINFIELD, Preston†	Sh	Stevens	*Athenaeum*, **1846**: 537
June 11	ANON	Birds Min Sh Ins Cor Lib	Stevens	*Athenaeum*, **1846**: 569
June 25-26	MOORE (Dr.), Preston, Lancashire	Fos Min Sh	Stevens	*Athenaeum*, **1846**: 569
June 29	ANON, Candler's Warehouse, Billiter Street.	Mam S.African hh	Lewis & Peat, *house sale*	*Athenaeum*, **1846**: 642
July 2-3	KNIGHT (R.), Tavistock Square†	Min Fos Sh Brazilian ins.	Stevens	*Athenaeum*, **1846**: 642
July 17	ANON	Sh Min Fos Ins Herb	Stevens	*Athenaeum*, **1846**: 697
July 30	ANON	Sh Min Bird skeletons Mam skeletons	Stevens	*Athenaeum*, **1846**: 745
August 26	WEBSTER (Thomas)†	Min Fos	Rushworth & Jarvis 247 lots 15pp.	BMNHP
October 26-30	BANKS (D.), Sheppy Court, near Sheerness, Isle of Sheppey, Kent.	26-28 x: Fos Sh Fish Cor Birds Quad Nh, 212	Attwater, *house sale* 952 (incl. 41 blank) + 32 bis lots 47pp.	ASH
1847				
February 12	WHITEHILL (Colonel)†	Indian ins.	Stevens	*Athenaeum*, **1847**: 113
March 15-16	HEULAND (Henry)	Min	Stevens 400 + 2 bis lots 24pp.	BMNHM f.p.f.n. (BM purchases only)
March 18	DURANT (Lt. Colonel)†	Sh	Stevens	*Athenaeum*, **1847**: 274
April 1	WEATHERHEAD†	Br ins.	Stevens	*Athenaeum*, **1847**: 322

1847 – 1848	SOURCE	CONTENTS	AUCTIONEER & SALE CAT.	REFERENCE
June 14-15	HEULAND (Henry)	Min	Stevens 402 + 3 bis lots 28pp.	BMNHM f.p.
July 22-23	SIRR (Major Henry Charles), Dublin†	Fos Min Sh	Stevens	*Athenaeum*, 1847: 753
November 2-6	HUGHES'S NATIONAL MAMMOTH ZOOLOGICAL & EQUESTRIAN ESTABLISHMENT, Royal Gardens, Vauxhall	4 xi: Mam, 111	Tattersall, *house sale* 770 lots 32pp.	B
November 6	'A Gentleman'	Bird skeletons Mam skeletons, 41	Stevens 220 lots 8pp.	RCS
December 13	FRASER (Louis), late Curator Zoological Society of London	For live quad. For live birds	Stevens	*Athenaeum*, 1847: 1234; Allingham (1924: 39)
December 14-15	DUNSTON (John)†	Min	Stevens 405 + 1 bis lots 22pp.	GEOL
1848 January 14	LONGLEY (Henry)†	Br ins. Entomological app.	Stevens	*Athenaeum*, 1848: 28; Allingham (1924: 33)
February 24-26	(a) ADAMSON (John), F.L.S., Newcastle-on-Tyne (b) ANON, 'unreserved stock of a dealer retiring from the trade'	Sh	Stevens 720 lots 31pp.	NMW p. add.
April 14	HEULAND (Henry)	Min Met	Stevens 200 lots 12pp.	BMNHM
May 5	ENDERBY (Mrs.), Blackheath†	Min	Stevens	*Athenaeum*, 1848: 449
May 12	ANON	Min Fos Sh Lib	Stevens	*Athenaeum*, 1848: 449
June 2-3	BOYS (Major), Woolwich	Min Fos Sh Br birds Rept	Stevens	*Athenaeum*, 1848: 521
June 10	ANON	Sh Birds Cor Min, 75	Puttick 404 + 2 lots 14pp.	BML p.n.
August 15-19, 21-25, 29-31, September 1-2, 4-8, 12-16, 18-22 26-30, October 3-7	GRENVILLE (Richard), 2nd Duke of Buckingham & Chandos, Stowe House, near Buckingham	3 x: Birds Rept Mam Sh Cor Fos Ins Min, 24	Christie, *house sale* 6301 lots (approx.) VIII + 271 + 10 + 28pp. + 2 plates	C p.n.
September 4-5	COTTON (Rev. Horace Salusbury), Newgate†	Sh Fos, 37	Sotheby 380 + 3 bis lots 19pp.	BML p.n.; BMP

1848 – 1850	SOURCE	CONTENTS	AUCTIONEER & SALE CAT.	REFERENCE
September 6-7	STUTCHBURY [natural history dealer]	Sh Min Fos (his stock of these groups	Stevens	*Athenaeum*, **1848**: 873
October 31	(a) RADDON (Wm.), Bideford† (b) BOYS (Captain W.J.E.)	(a) Br ins. For ins (b) Indian ins.	Stevens	(a) *Athenaeum*, **1848**: 1041 (b) *Athenaeum*, **1848**: 28; Allingham (1924: 33)
December 8	(a) 'A late well-known collector in Norfolk'[1] (b) TUCKER, Quadrant, Regent Street	(a) Br lep Col Orth Hym Neur (b) Stock of exotic ins.	Stevens 211 lots 12pp.	J p.n.
1849 February 1-2	(a) MILLER (Thomas), Plymouth† (b) ANON	(a) Land sh Marine sh (b) Birds Fos Min Rept Herb	Stevens	*Athenaeum*, **1849**: 82
February 6	(a) ANON (b) 'A Gentleman'	(a) Ins (b) Br lep (coll. 'formed by a gentleman at Brighton')	Stevens	*Athenaeum*, **1849**: 82
May 21-24	FORSTER (Edward), Vice President Linnaean Society, Woodford, Essex.	Lib Herb Fos Nh	Sotheby 1264 lots 68pp.	**BMNHB**
June 19	ANON	Sh Fos Min Cab Irr	Stevens	*Athenaeum*, **1849**: 610
June 22	BOYS (Captain W.J.E.)	Indian col.	Stevens	*Athenaeum*, **1849**: 610
June 28	ANON	Fos	Stevens 176 + 1 bis lots 8pp.	**BMNHP**
1850 March 27	TOMKINS (John), Ock Street, Abingdon†	Br birds ('rare British birds in 60 glass cases')	Harris & Belcher, *house sale* 60 lots 2pp.	**CAMB**
April 26-27	ROSS (General)†	Sh	Stevens	*Athenaeum*, **1850**: 411
April 30	ROWE (Richard)	Nh, 2	Christie 91 lots 6pp.	C m.p.; BUP
May 3	(a) WARSZEWEIZ (A.) (b) ANON	(a) Col. (b) Chinese ins.	Stevens	*Athenaeum*, **1850**: 435
June 7	(a) DOUBLEDAY (Edward)† (b) [HEWITSON]	(a) Lib (b) Br ins, 54	Stevens 247 + 2 bis lots 12pp.	H s.p. (Doubleday lots only)
August 7-9	KIRBY (Rev. Wm.), Rectory House, Barham near Ipswich†	Lib, 120	Garrod, *house sale* 548 lots 34pp.	H f.p.; **BMNHL** copy

[1] Asterisked specimens are referred to in C.J. & J.Paget's *Natural History of Yarmouth*. Also included were 24 lots of Scottish lep.

1850–1852	SOURCE	CONTENTS	AUCTIONEER & SALE CAT.	REFERENCE
August 16	(a) JERDON (T.C.) (b) ANON	(a) Birds Anim (b) Ins Min Fos Sh	Stevens	*Athenaeum*, **1850**: 826
September 22	DAVIES (George), Scarborough†	Min Fos, 23	Sotheby 147 + 3 bis lots 12pp.	**BML** p.n.; **BMM** p.n.
1851 January 15-18	GUBBA (A.L.), 'late of Havre de Grace'	Exotic sh	Stevens 817 lots 34pp.	**CAMMUS; SPD** copy; **CAMB** f.p. **BMNHL** f.p. copy
May 27-28	BLAND (Michael)†	Exotic sh Lib	Stevens	*Athenaeum*, May **1851**; 538 Allingham (1924:34)
July 11	(a) GARDNER (George), Director Royal Botanic Garden, Ceylon† (b) ANON	(a) Bot lib Nh lib, 200 (b) Min, 64	Stevens 264 lots 12pp.	**BMNHB**
July 29-30	CUTTING (Dr.), Barbados	Sh Min Fos	Stevens	*Athenaeum*, June **1851**: 730, 762
September 4	(a) AITON (J. Townsend), Kensington Palace Avenue (b) SCHLEICKER (M.), Bex, Switzerland (c) TEESDALE (d) ANON	(a) Bot lib, 247 (b) Herb, 1 (c) Herb, 1 (d) Min Sh Irr	Foster 250 + 2 bis lots 9pp.	**BMNHM; VA** p.n.
October 6-11	STANLEY (Edward Smith, 13th Earl of Derby), Knowsley Hall, near Liverpool†	Menagerie Aviary	Stevens, *house sale* 641 lots 4 + 50pp.	**BMNHZ** f.p. add (88 lots) **LIV; CAMB** p.n.
December 10	PHILLIPS (Richard)	Lib Met Min	Sotheby 272 lots 14pp.	**BML** p.n.; **S** p.n. copy
December 10-15	DUNN (Nathan)	11 xii: Nh, 14	Christie 600 + 3 bis + 18 lots 35pp.	**C** m.p. add; **BMP**
1852 January 9	ANON	Sh Min Nh Irr	Stevens	*Athenaeum*, **1852**: 34
January 28	ANON	Birds, 1	Foster 122 lots 8pp.	**VA** p.n.; **EKW**
January 30	ANON	Sh Min Fos Birds Herb Irr	Stevens	*Athenaeum*, **1852**: 97
February 27-28	(a) MURRAY (Dr.), Hull† (b) [ANON]	(a) Lib (b) Min Fos Sh Herb	Stevens	*Athenaeum*, **1852**: 186
April 2	ANON	Lep Birds Cab Animals Min Fos	Stevens	*Athenaeum*, **1852**: 338
April 6	(a) MILTON (b) [ANON]	(a) Br birds (b) Ool	Stevens	*Athenaeum*, **1852**: 314

1852 – 1853	SOURCE	CONTENTS	AUCTIONEER & SALE CAT.	REFERENCE
April 23	ANON	Fos Min Br birds For birds Lib Nh	Stevens	*Athenaeum*, **1852**: 419
May 4	COLE (J.W.)	Min Fos	Stevens	*Athenaeum*, **1852**: 475
May 7	ANON	Birds Min Nh	Stevens	*Athenaeum*, **1852**: 475
May 21	(a) – (b) ANON	(a) Birds (b) Min Fos Sh	Stevens	*Athenaeum*, **1852**: 531
June 4	ANON	Min Birds Nh Irr	Stevens	*Athenaeum*, **1852**: 594
June 18	ANON	Sh Min Ins Birds Cab	Stevens	*Athenaeum*, **1852**: 642
July 30	ANON	Rept Sh Birds Animals Nh	Stevens	*Athenaeum*, **1852**: 786
December n.d.	BIRMINGHAM PHILOSOPHICAL INSTITUTION, Canon Street, Birmingham[1]	'Geological and Mineralogical Museum'	Hornblower, *house sale*	*Athenaeum*, **1852**: 1282
December 3	(a) FISHER (Miss), Westcott, Dorking (b) 'A well-known amateur' (c) ANON	(a) Br birds Br ool (b) Dup ool (c) Br birds	Stevens	*Athenaeum*, **1852**: 1282
December 17	ANON	Min Fos Nh Irr	Stevens	*Athenaeum*, **1852**: 1346
1853 n.d.	STOTHARD (Thomas), artist	Butts	Stevens	Allingham (1924:33, 145)
n.d.	ANON., incl. the Laurens, Theodorus GRONOVIUS coll. of fish	Fish Irr	Phillips	Wheeler *Bull. Brit. Mus. nat. Hist.* (hist.ser.) **1** (5), 1958:193
January 26	WOLLEY (John, jun.)	Lapland ool	Stevens	*Athenaeum*, **1853**: 66
February 8	ANON	Br ins. Sh Fos Irr	Stevens 211 lots 8pp.	J
February 11	WISE (J.R.)	Ool	Stevens	Wolley, *Ootheca Wolleyana*, **3**:86
May 2-4	MANTELL (G.A.), L.L.D., F.R.S., 19 Chester Square, Pimlico	Fos Min Lib, 349	Foster, *house sale* 595 lots 40pp.	T p.; **BMNHL** p. copy

[1] It is possible the sale was abandoned and that this museum was presented to Queen's College, Birmingham (cf. Aris's *Birmingham Gazette,* 29 xi 1852 & 13 xii 1852).

1853–1854	SOURCE	CONTENTS	AUCTIONEER & SALE CAT.	REFERENCE
May 24	POTTS (Thomas H.), Kingswood Lodge, Croydon	Ool (incl. 2 Great Auk eggs)	Stevens	Parkin (1911:6 Allingham (1924:34, 159 162)
June 6	HURT (Charles)†	Min Fos Cab, 116	Sotheby 122 lots 9pp.	BML p.n.; S p.n. copy
June 9-13, 15	ANON	13 vi: Min, 1	Sotheby 1072 + 4 bis lots 67pp.	BMM p.n.; BML
1854 February 17	WOLLEY (John, jun.)	Lapland ool	Stevens	*Athenaeum*, **1854**: 166
April 7	ANON [? T.H.POTTS]	Ool (incl. Great Auk egg, lot 101)	Stevens	Parkin (1911:6 who refers to a cat. in Alfred Newton's library
May 9	TRISTRAM (Rev. H.B.)	Ool	Stevens	*Athenaeum*, **1854**: 510
May 30-31	STOKES (Charles), F.R.S., F.S.A.†	Lib (much bot.)	Sotheby 751 lots 44pp.	**BMNHB**
June 1	STOKES (Charles), F.R.S., F.S.A.†	Min Fos Sh Cor Ins Nh, 136	Sotheby 165 lots 9pp.	BML p.n.; S p.n. copy
June 26-27	TAYLOR (Henry)†	Fos	Stevens 335 lots 15pp.	**BMNHP; GEOL**
June 30	(a) FOLKES (W.) (b) SMITH (W.P.)	(a) Br ool (b) Fish Rept Sh	Stevens	*Athenaeum*, **1854**:735
July 21-22	[HURFORD]	Sh	Stevens 410 lots 16pp.	CAMB m.p.; **BMNHL** m.p. copy
July 26-28, 31– August 4, 7-11 14	HOLFORD (James), Holford House, London†	3 viii: Ins Sh Nh, 3	Foster, *house sale* 2033 (incl. 61 blank) + 1 bis lots 99pp.	EKW; **BMNHL** copy
August 4	(a) CONDAMINE (Rev. H. de la), Blackheath (b) ANON	(a) Fos Min Sh Nh (b) Birds Lib	Stevens	*Athenaeum*, **1854**: 926
October 11	JEPHSON (Dr.), incl. the EDINGTON cabinet of shells	Min Sh	Stevens	*Athenaeum*, **1854**: 1159
October 24	ANON	Birds Ool	Stevens 271 lots 10pp.	**BMNHZ**
October 25	TRUMAN (J.), Edwinstowe, Notts.†	Br ins (incl. 'many rarities and others peculiar to Sherwood Forest')	Stevens	*Athenaeum*, **1854**: 1223

1854–1855	SOURCE	CONTENTS	AUCTIONEER & SALE CAT.	REFERENCE
November 10	ANON	Min Fos Sh Rept Ins Birds Nh Irr	Stevens 252 + 1 bis lots 10pp.	H inc. (lacks pp.5-6, lots 56-124)
2 ffdd	COCKBURN (Lord)†	27 xi: Stones, 1	Nisbet, *Edinburgh*	GLA
December 8	ANON	Sh Min Bot Fos Irr	Stevens 249 lots 10pp.	CAMB m.p.; **BMNHL** m.p. copy
1855 N.d.	SURREY ZOOLOGICAL GARDENS, Kennington		Stevens	Allingham (1924:39)
January 26	WOLLEY (John, jun.)	Lapland ool	Stevens 206 lots 12pp.	**BMNHZ**; N m.p.
January 27	ANON	Birds Ool	Stevens 240 lots 10pp.	**BMNHZ**
March 16	(a) – (c) ANON	(a) Min Sh (b) Birds Nh (c) Nh Irr	Stevens 236 lots 10pp.	CAMB; **BMNHL** copy
April 10-11	PROSSER (R.), Part III	Min, 2	Chesshire & Gibson, *Birmingham* 714 lots 28pp.	BIRM
April 26-27	'A Nobleman'† [J. da AMIS...]	Sh	Stevens 539 lots 22pp.	CAMB m.p., **BMNHL** m.p. copy
April 28	'A Nobleman'† [J. da AMIS...]	Sh	Stevens	Advert on title p. of Stevens' cat. of 26-27 IV 1855
May 4	ANON	Birds Ool	Stevens 256 lots 12pp.	**BMNHZ**
May 18	ANON	Ool Mam Ins	Stevens 243 lots 10pp.	**BMNHZ**
June 1	HASTINGS (Marchioness of)	Fos	Stevens 256 + 1 bis lots 10pp.	**GEOL**
June 22	ANON	Lep Birds Sh Mol Nh Irr	[Stevens] 165 lots [pp. 3-8]	H inc
July 5	(a) FYSH (Rev. F.), Torquay (b) ANON	(a) Sh (b) Min Birds HH	Stevens	*Athenaeum*, 1855: 747
July 27	WEAVER (Thos.)	Min Fos	Stevens	*Athenaeum*, 1855: 827
November 2	(a) – (c) ANON	(a) Ool (b) Ins (c) Birds Nh	Stevens 288 + 2 bis lots 12pp.	H

1855–1857	SOURCE	CONTENTS	AUCTIONEER & SALE CAT.	REFERENCE
November 9	ANON†	Ool	Stevens 239 lots 10pp.	BMNHZ
December 13	ANON	Fos Col Nh, 14	Sotheby 266 lots 14pp.	BML p.n.; S p.n. copy
1856 January 25	(a) BACON (Dr.)† (b) ANON	(a) Sh (b) Min Fos	Stevens 235 lots 10pp.	BMNHZ
January 29	HORTICULTURAL SOCIETY OF LONDON	Herb	Stevens 54 + 3 bis lots 4pp.	BMNHB p.n. add.
February 5-6	HIGHLEY, jun., 'retiring from business'	Min Rept Animals Lib App Nh	Sotheby 570 + 1 bis lots 2 + 26pp.	GEOL
February 12	(a) WING (William) (b) ANON	(a) Br ins (b) Borneo col.	Stevens	*Athenaeum*, Nov. **1856**: 155
March 7	WOLLEY (John, jun.)	Ool	Stevens 201 lots 20pp.	BMNHZ; N s.p.
April 15	ANON	Sh	Stevens	*Athenaeum*, April, **1856**:44
April 18	(a) 'A well-known collector' (b) ANON	(a) Br & European lep Br & European col. (b) Ool	Stevens [lots 1-141] [pp. 1-8]	H inc.
May 2-3	MUNN (Henry)†	Sh Min	Stevens 235 lots 12pp.	NMW inc. (lacks pp. 5-8, lots 37-41)
May 8-9	[FITTON (Dr.)]	Lib	Sotheby 667 + 43 bis lots 2 + 40pp.	GEOL
May 29-31 June 2	IMAGE (Rev. Thos.) Whepstead Rectory, near Bury St. Edmunds	29 v: Fos Sh Min Nh, 371	Newson, *house sale* 1441 (incl. 78 blank) + 1 bis lots 62pp.	GEOL
June 20	PEREIRA (Dr.)	Min Fos	Stevens	*Athenaeum*, June **1856**: 732
November 11	LAMB (Charles)	Br ins For ins	Stevens 226 + 2 bis lots 12pp.	J
December 4-6	YARRELL (William)†	Br birds Ool Fish Nh Irr	Stevens 522 lots 19pp.	RCS
1857 January 26-27	BUCKLAND (Rev. Dr.), Dean of Westminster†	Lib (much geological)	Stevens 501 lots 24pp.	GEOL

1857 – 1858	SOURCE	CONTENTS	AUCTIONEER & SALE CAT.	REFERENCE
January 30	BUCKLAND (Rev. Dr.), Dean of Westminster†	Min Fos Nh Irr	Stevens	J.H.O. Burgess, *Eccentric Ark: the Curious World of Frank Buckland* (1968:68)
February 9	BOTANICAL SOCIETY	Lib Herb	Stevens 251 + 2 bis lots 10pp.	**BMNHB**
February 10	TRISTRAM (Rev. H.B.)	Ool (collected in Algeria in 1856)	Stevens 171 lots 15pp.	**BMNHZ; N** p.; JAG
February 20	ANON	Birds Ool	Stevens 251 lots 12pp.	**BMNHZ**
March 13	(a) HEMMINGS (J.), Brighton† (b) ANON	(a) Br ins. (b) Br ins Exotic ins	Stevens	*Athenaeum*, March, **1857**: 295
April 24	SOWERBY (G.B.)	Sh Min Fos (his stock of these groups)	Stevens	*Athenaeum*, April, **1857**: 488
May 12	WOLLEY (John, jun.)	Ool	Stevens 203 lots 16pp.	**BMNHZ** p.; N p.
May 29	EDWARDS, 'late of Camden Town'	Min Sh (his stock of these groups)	Stevens	*Athenaeum*, May, **1857**: 647
July 24	BAKER (John), Cambridge	Ool	Stevens	*Athenaeum*, July, **1857**: 895
November 24-25	KEATE (R.), 11 Hertford Street, Mayfair†	25 xi: Birds Min Sh Butts, 3	Oxenham, *house sale* 444 lots 28pp.	**RCS**
1858 January 26	PROCTOR (William), Durham University Museum	Ool	Stevens 199 lots 8pp.	**BMNHZ; N**
February 9	TRISTRAM (Rev. H.B.)	Ool	Stevens 291 lots 19pp.	**BMNHZ; N** p.; JAG
February 12	INGALL (Henry)	Ins Nh	Stevens	*Athenaeum*, Feb., **1858**: 162
February 23	WOLLEY (John, jun.)	Ool	Stevens 212 lots 24pp.	**BMNHZ** p.; N p.
March 19	ANON	Min Sh Fos Birds Nh	Stevens	*Athenaeum*, March, **1858**: 322

1858–1859	SOURCE	CONTENTS	AUCTIONEER & SALE CAT.	REFERENCE
March 23	PRESTON (Edward S.)	Birds Ool	Stevens 193 lots 8pp.	**BMNHZ**; N m. add (ms. notes by E.S.Preston
April 14	(a) GOURLIE (Wm.), Glasgow† (b) ANON	(a) Herb, 77 (b) Pine cones & seeds, 68	Stevens 137 + 8 bis lots 7pp.	**BMNHB**
April 16-17	ENTOMOLOGICAL SOCIETY OF LONDON	Exotic ins Cab	Stevens 368 lots 12pp.	J; (RES)
May 14	WILLIAMS, Oxford Street†	Birds (his stock of)	Stevens	*Athenaeum*, May, **1858**: 579
May 25	GRIFFITH (Edward)†	Fos	Stevens	*Athenaeum*, May, **1858**: 579
July 12 ffdd	HEYSHAM (T.C.), Carlisle†	Lib	Stevens 2984 lots 74pp.	**RCS**
November 9-10	(a) BROWNELL (George), Liverpool† (b) 'A Gentleman'	(a) Br lep (b) Lep Col.	Stevens	*Athenaeum*, Nov. **1858**: 571
1859 January 14	STREATFIELD (J.F.)	Br ins	Stevens	*Athenaeum*, Jan. **1859**: 35
March 8	WOLLEY (John, jun.)	Ool	Stevens 182 lots 16pp.	**BMNHZ**; N m.p
April 27	MARTIN (K.B.), Harbour Master	Fos Min	Hinds, *Ramsgate*	*Athenaeum*, April, **1859**: 536
April 28-30	ASHMEAD (G.B.), Grosvenor Square	Birds Rept Animals (his stock of these groups)	Stevens	*Athenaeum*, April, **1859**: 503, 535
May 2-5	HORTICULTURAL SOCIETY (The)	Lib Bot drawings	Sotheby 985 lots 50pp.	**BMNHB; BML** p.n.
May 11	HEYSHAM (T.C.)	Br birds	Stevens	*Athenaeum*, May, **1859**: 599
May 13	HEYSHAM (T.C.), Carlisle†	Br ins (incl. Lep Col Hym Dipt Neur)	Stevens 164 lots 10pp.	J
May 16	HEYSHAM (T.C.)	Ool	Stevens	Wolley, *Ootheca Wolleyana*, 2: 270, 3:111
May 27	HEYSHAM (T.C.) Carlisle†	Birds Sh Min Fos	Stevens 213 lots 8pp.	**RCS**

1859 – 1860	SOURCE	CONTENTS	AUCTIONEER & SALE CAT.	REFERENCE
ne 27-29	BRODERIP (W.J.), F.R.S.†	Lib	Sotheby 702 + 3 bis lots 2 + 53pp.	**GEOL**
ne 27-29	BROWN (Robert), D.C.L., F.R.S.	Lib	Stevens 793 lots 34pp.	**RCS; GEOL** inc.
ly 26	BROWN (Robert)†	Herb	Stevens	*Athenaeum*, July, **1859**: 99
ly 29	(a) LEE (William) (b) ANON	(a) Fos (b) Ool Nh	Stevens 212 lots 15pp.	**RCS; GEOL**
ctober 21	(a) ANON (b) 'A nobleman'	(a) Herb (b) Herb Lib	Stevens	*Athenaeum*, Oct. **1859**: 452
ovember 11	ANON	Birds	Stevens	*Athenaeum*, Nov. **1859**: 582
ovember 22	HORSFIELD (Dr. T.)	Lib	Stevens	*Athenaeum*, Nov. **1859**: 619
ovember 23-24	(a) HORSFIELD (Dr. Thomas), F.R.S., Librarian East India Company† (b) 'A Gentleman'	(a) Ins (b) Br col For col Br lep	Stevens 388 + 3 bis lots 18pp.	**J; H; BMNHL** copy
ovember 25	HORSFIELD (Dr. Thomas), F.R.S., Librarian East India Company	Herb Min Lib	Stevens 219 (lots 389-607) 12pp.	**RCS**
ecember 14-17	VERNEDE (H.), West End, Hampstead†	Sh	Stevens 814 lots 34pp.	**CAMB** m.p.; **BMNHL** m.p. copy
ecember 22	LAUTOUR (Albert de)	Br birds For birds	Jackson, *Hitchin*	*Athenaeum*, Dec. **1859**: 759
1860 ebruary 11	(a) – (b) ANON	(a) Lep Col (b) Exotic lep Exotic col.	Stevens 228 lots 12pp.	**J**
arch 13	DIXON (Rear Admiral M.H.)	Sh	Stevens	*Athenaeum*, March **1860**: 287, 323
arch 15-16	NUTTALL (Dr.), Professor of Botany, University College, Massachusetts†	Min (coll. formed 'during a long residence in North America')	Stevens 485 lots 24pp.	**GEOL**
arch 27	TRISTRAM (Rev. H.B.)	Birds	Stevens 217 lots 16pp.	**JAG; CH** copy
pril 24	ANON	For birds For animals	Stevens	*Athenaeum*, April, **1860**:459
ay 8-9	STUTCHBURY (Samuel), Curator Bristol Institution, Government Surveyor, Australia†	Sh Min Fos Lib Nh	Stevens 371 lots 15pp.	**GEOL**

1860–1861	SOURCE	CONTENTS	AUCTIONEER & SALE CAT.	REFERENCE
May 22	(a) ANON (b) STUTCHBURY (S.)†, (c) – (d) ANON	(a) Br lep (b) Australian ins (c) Indian ins (d) Br lep	Stevens 187 + 6 bis lots 12pp.	J
May 30-31	WOLLEY (John, jun.)	Dup ool	Stevens 376 lots 27pp.	BMNHZ; N; WHB
October 19	BROUGHTON (Edward), Edinburgh†	Min	Stevens	*Athenaeum*, Oct. **1860**:467
October 23-24	(a) ARMSTRONG (Dr.)† (b) ROYAL UNITED SERVICES INSTITUTE (c) FINCH (Lady)	(a) Min Fos Sh (b) Birds Min Sh HH (c) Sh	Stevens	*Athenaeum*, Oct. **1860**:467
November 20-21	LYNCH (Christopher), 2 Sumner Place, Onslow Square, SW.†	21 xi: Min Fos Sh HH Nh, 136	Smith 584 (incl. 24 blank)+ 12 bis lots 43pp.	BML
December 7-8	(a) BURMAN (R.) (b) HUBBARD (J.), Bury St. Edmunds	(a) Mauritius sh (b) Fos Sh	Stevens	*Athenaeum*, Dec. **1860**: 731
December 14-15	(a) BRIGHT (John)† (b) – (c) ANON (d) ROYAL UNITED SERVICES INSTITUTE	(a) Birds (b) Birds HH (c) Ins (d) Nh	Stevens	*Athenaeum*, Dec. **1860**:731
1861 January 18	BURGESS (S.), Westbrook, Lydd†	Fos Br birds	Ronald & Buss, *Ashford, Kent*	*Athenaeum*, Jan. **1861**:35
January 18	ANON	'Cryptogamic & Dried Plants'	Stevens	*Athenaeum*, Jan. **1861**:35
February 1	(a) BOUSFIELD (W.S.), Dulwich (b) ANON	(a) Br ool (b) Br lep.	Stevens	*Athenaeum*, Feb. **1861**:138
February 15	(a) SARGENT (Frederick) (b) ANON	(a) Min (b) Fos Sh	Stevens	*Athenaeum*, Feb. **1861**:175
March 8	(a) BREWIN (S.)† (b) ANON	(a) Sh (b) Chinese ins	Stevens	*Athenaeum*, March **1861**:27
March 15	WHEELWRIGHT (H.W.)	Br ool For ool	Stevens	*Athenaeum*, March **1861**:27
April 23-24	[SALMON (J.D.)]	Ool	Stevens	*Ibis*, **1863**:371-372
May 23	(a) – (f) ANON	(a) Br lep (b) Br lep (c) Br col (d) Br hym (e) For ins (f) Br lep	Stevens 214 lots 12pp.	J
June 14	ARMFIELD'S MACCLESFIELD MUSEUM	Birds Animals	Stevens	*Athenaeum*, May **1861**:680, June **1861**:747
July 1-2	HENSLOW (Rev. Prof.), Hickam Rectory, Suffolk.	Lib	Stevens	*Athenaeum*, May **1861**:680
July 9	WALKER (Edwin), Enfield†	Min	Stevens	*Athenaeum*, July **1861**:3

1861–1863	SOURCE	CONTENTS	AUCTIONEER & SALE CAT.	REFERENCE
ly 19	ANON	Sh	Stevens	*Athenaeum*, July **1861**:35
ly 23	HENSLOW (Rev. Prof.)	Fos Sh Min Ins Birds Quad Nh, 190	Stevens 263 lots 12pp.	**GEOL**
ctober 8	SHEFFIELD†	Min	Stevens	*Athenaeum*, Sept. **1861**:395
ovember 21	(a) WHEELWRIGHT (H.W.) (b) ANON	(a) Swedish ool (b) N. American birds	Stevens	*Athenaeum*, Nov. **1861**:599, 600
ovember 29	BELL (Prof.)	Fos Min Birds Rept	Stevens	*Athenaeum*, Nov. **1861**:670
cember 10-14	QUEKETT (Prof. John Thomas), F.R.S., Conservator of Hunterian Museum	10 xii: Lib. 11 xii: Microscopes 12 xii: Fos Min Sh Ins Nh, 20	Bullock 932 + 24 bis lots 45pp.	**GEOL**
62				
rch 7	LIMMINGHE (Comte Alfred de), Brussels†	Herb Fungi	Stevens 103 + 4 bis lots 8pp.	**BMNHB**
rch 25	(a) BENTLEY (William)† (b) ANON (c) TURNER (J.A.) (d) ANON	(a) Br ins Ool (b) Lep (c) Col (d) Swedish birds	Stevens 247 lots 12pp.	**CH; J**
ril 9	WOLLEY (John, jun.)†	Lapland ool	Stevens 163 lots 13pp.	**BMNHZ; WHB; N**
y 23	PHILLIPS (Rev. E. J. March), Stathern	Min Fos	Stevens	*Athenaeum*, May **1862**:647
y 30	(a) AUCKLAND (J.T.), Eastbourne (b) ANON	(a) Sh Conchological lib (b) Herb Animals	Stevens	*Athenaeum*, May **1862**:678
ly 1	(a) ATKIN (Dr. George), Hull† (b) – (c) ANON	(a) Herb (b) Herb, 74 (c) Min Fos Sh, 129	Stevens 243 + 5 bis lots 12pp.	**BMNHB**
ly 11	HAWKINS (William), F.S.A., Kensington†	Mam Birds Min Ins Rept Nh, 12	Sotheby 262 + 1 bis lots 16pp.	BML p.n.; **S** p.n. copy
ctober 28-30	CLARKE (Robert Mayne), Cold Harbour Mansion, near Wallingford, Berkshire	28 x: Br birds For birds Rept Fish Ool Sh Fos Ins Seaweeds Fungi	Mallam, *house sale*	*Athenaeum*, Oct. **1862**:482
ovember 20-21	WHEELWRIGHT (H.W.)	Lapland birds Lapland animals	Stevens	*Athenaeum*, Nov. **1862**:580
63				
rch 24	(a) WALTON (John) (b) – (d) ANON	(a) Ins (b) Br col (c) Br lep (d) Br col	Stevens 256 lots 16pp.	**J** m.n.
y 6-7	BARRY (Dr. Martin)	Ool	Stevens	*Ibis*, **1863**:372, 477-478

1863 – 1864	SOURCE	CONTENTS	AUCTIONEER & SALE CAT.	REFERENCE
May 26	INGALL (Thomas)†	Fos Sh Ool Br birds Br lep Nh, 262	Sotheby 287 + 1 bis lots 12pp.	BML p.n.
June 8	CURTIS (John)†	Entomological lib	Stevens	*Athenaeum*, June **1863**:731
June 12	(a) DU CHAILLU (Paul) (b) ANON	(a) Birds Mam (b) Birds Mam	Stevens 255 lots 11pp.	**BMNHZ; RCS**
June 13	(a) – (b) ANON (c) SAULL†	(a) Sh (b) Min (c) Fos	Stevens	*Athenaeum*, June **1863**:731
July 8	(a) ENTOMOLOGICAL SOCIETY OF LONDON (b) ANON (c) CURTIS (John)† (d) – (e) ANON (f) ARMFIELD, Macclesfield (g) ANON	(a) Br ins For ins (b) Exotic ins (c) Ins (d) Australian ins (e) Exotic butts (f) Br lep (g) Lib	Stevens 217 lots 12pp.	**J** p. of (a); **H**
July 9	ANON	Min	Stevens	*Athenaeum*, July **1863**:3
July 27	INGALL (Thomas)†, incl. the coll. 'of the late Mr. HATCHETT'	Br ins, 170	Sotheby 193 lots 12pp.	BML p.n.; **S** p.n. copy
August 13-14	SHUCKARD (W.E.)	Lib Ins Irr	Sotheby 742 lots 46pp.	**J**
October 13	PAMPLIN (Wm.), Frith St., Soho	Bot lib	Stevens	*Athenaeum*, Oct. **1863**: 451
November 10	(a) LINNEAN SOCIETY (b) PULTENEY (Dr.) (c) ANON	(a) Ins (b) Sh Fos (c) Ins Birds HH Bot Nh	Stevens 268 lots 14pp.	**BMNHB; H;** LINN p. add; **SPD** p. add. cop
1864 January 12	F.......... (R.T.)	Br ool	Stevens 228 lots 10pp.	**BMNHZ**
February 9	(a) LORD (J.K.) (b) ANON	(a) Birds Animals (b) Birds Ool	Stevens	*Athenaeum*, Nov. **1864**:178
April 12	(a) EVANS (Herbert N.), Hampstead (b) ANON	(a) Ool (b) Ool from Turkey & Canada	Stevens	*Athenaeum*, April **1864**:491
May 6	WATSON (John)	Birds Mam Rept Min (all these groups from Queensland)	Stevens	*Athenaeum*, April **1864**:594
May 19	WOLLEY (John, jun.)†	Lapland ool	Stevens 185 lots 14pp.	**N**
May 24-25	REEVE (Lovell), incl. William METCALFE, Dr. GASKOIN, Dr. KRAPP	Sh	Stevens	*Athenaeum*, May **1864**:630
September 2	BURGON (John Towry)	Fos Nh	Stevens	*Athenaeum*, Aug. **1864**:259

1864 – 1866	SOURCE	CONTENTS	AUCTIONEER & SALE CAT.	REFERENCE
ovember 8	(a) RUSSELL (W.T.), Ringwood† (b) REID (H.), Doncaster	(a) Br lep (b) Br lep	Stevens	*Entomologist*, 2: 104,120
ovember 22	WHEELWRIGHT	Ool (from Sweden & Lapland, 1863, 1864) Birds	Stevens 381 lots 16pp.	**BMNHZ**
65 d.	CLARK (Rev. Hamlet)	Br lep Br col	Stevens	Horne & Kahle (1935-37:43); Allingham (1924:151)
d.	GILL (Dr. Battershell)	Br lep	Stevens	Horne & Kahle (1935-7:90)
d.	PREST (William)	Br lep	Stevens	Horne & Kahle (1935-37:215)
bruary 17	IRVING (Lt.Gen.) Balmoral House, Kirkcudbright†	Sh	Stevens	*Athenaeum*, Feb. **1865**:183
arch 2	(a) FRASER of LOVAT (Hon. Archibald)† (b) ANON	(a) Fos Min Sh, 106 (b) Min Fos Sh Ins, 21	Sotheby 290 + 1 bis lots 18pp.	**BML** p.n.
arch 7	FRASER OF LOVAT (Hon. Archibald)†	Min Fos Nh, 5	Sotheby 227 + 1 bis lots 16pp.	**BML** p.n.
arch 28	(a) READ (Rev. G. Rudston) (b) ANON (c) BOUCHARD (Peter) (d) ANON	(a) Br lep (b) Br lep (c) Br ins (d) For ins	Stevens 224 lots 12pp.	**J** p.m.n.
pril 24-29	DENNISON (J.)†	Sh Cab Conchological lib	Stevens 1206 lots 58pp.	**J** p.; **CAMB** m.p. **BMNHL** m.p. copy
ly 11	(a) TRISTRAM (Rev. H.B.) (b) WHEELWRIGHT (H.) (c) KRUPER (Dr.) (d) ANON	(a) Ool Birds (b) Ool Birds (d) Ool (incl. four eggs of the Great Auk)	Stevens 269 lots 14pp.	**RCS** inc. (pp.1-2 7-8, 13-14 only)
ctober 31	(a) ANON (b) SCHOMBURGK (Sir R.)† (c) ANON (d) BAIKIE (Dr.) (e) ANON	(a) Sh (b) Min Sh Siamese & Cambodian hh (c) HH Min (d) Birds (e) Fos	Stevens 281 lots 10pp.	**CAMB** s.p.; **BMNHL** s.p. copy
ovember 27 – ecember 1	BOWERBANK (Dr. J.S.), 20 Highbury Grove, Islington	Fos	Stevens 1189 lots 67pp.	**BMNHP**
ecember 5	BURCHELL (W.J.), D.C.L., Churchfield House, King's Road, Fulham.	Lib (mostly botanical, some irr.)	Foster 315 + 5 bis lots 16pp.	**H**
ecember 15	CUMING (Hugh)†, Part I	Dup sh	Stevens 250 lots 10pp.	**CAMB** m.p.; **BMNHL** m.p. copy
866 .d. ('early in e spring')	LINDLEY (Dr.), Phd., F.R.S., F.L.S.†	Lib (botanical)	Stevens	*Athenaeum*, Jan. **1866**:35

1866	SOURCE	CONTENTS	AUCTIONEER & SALE CAT.	REFERENCE
January 23	(a) REEVE (Lovell)† (b) ANON	Sh	Stevens 240 + 6 bis lots 10pp.	CAMB m.p.; BMNHL m.p. copy
February 9	REEVE (Lovell)†, F.L.S.	Conchological botanical & scientific lib	Stevens 221 + 5 bis lots 12pp.	CAMB m.p.; BMNHL m.p. copy
March 13	ANON	'Bones of the Dodo, from Mauritius'	Stevens 8 lots 1p.	BMNHP; N s.p. WHB
March 13	WHEELWRIGHT (H.)†	Ool Birds	Stevens	*Athenaeum*, March 1866:28
March 27	ANON	Birds Min Mam HH	Stevens	*Athenaeum*, March 1866:35
May 4	ANON	Sh Mam Exotic lep Nh	Stevens 214 lots 8pp.	BMNHZ
May 8	SARGENT (F.)	Min Fos	Stevens	*Athenaeum*, April 1866:547 Allingham (1924:55)
May 9	'A Gentleman... late Resident at Mauritius',† Part I	Sh	Stevens 274 + 9 bis lots 10pp.	NMW; CAMB m.p.; BMNHL m.p. copy
June 1-2	CUMMING (R. Gordon)†	Mam HH	Stevens 583 lots 16pp.	RCS s.p.; BMNHZ; CAMMUS m.p.
June 4	CUMMING (R. Gordon)†	Br ool Exotic sh	Stevens 324 lots 10pp.	BMNHZ
June 26-27	CUMING (Hugh)†	Dup sh Cab Lib	Stevens 418 lots 16pp.	H; BMNHL copy CAMB m.p. BMNHL m.p. copy
July 3	ANON ('several small private collections')	Sh Min	Stevens 343 + 1 bis lots 14pp.	CAMB f.p.; BMNHL f.p. copy
July 4	(a) CORBETT (B,), Piccadilly (b) SARGENT (F.)	(a) Birds Mam HH Nh (b) Br birds	Stevens	*Athenaeum*, June 1866:850
August 3	(a) GOODHALL (Henry Humphreys)† (b) ANON	(a) Fos Min (b) Min Sh	Stevens 249 lots 10pp.	BMNHM s.p.
December 6-7	'A Gentleman... late Resident at Mauritius'†, Part II	Sh	Stevens 507 lots 18pp.	CAMB p. add; BMNHL p. add copy
December 28	SEALY (A.F.), Cambridge	Br birds	Stevens	*Athenaeum*, Nov. 1866:663

1867 – 1868	SOURCE	CONTENTS	AUCTIONEER & SALE CAT.	REFERENCE
1867				
March 12	(a) FEATHERSTONHAUGH (G.W.)[1], 'formerly H.M. Commissioner for the Boundary between the United States and British North America'† (b) COLES (Henry), F.G.S.† (c) ANON	(a) Min (b) Fos (c) Min Fos Bones of Megatherium	Stevens 234 (incl. 5 blank) + 43 bis lots 10pp.	**BMNHM** m.p. f.n.
April 5	(a) DUFF (Richard) (b) DRESSER (H.E.)	(a) Birds Ool (b) Dup ool from Lapland, 1866	Stevens	*Athenaeum*, March **1867**:404
April 23	ANON†	Sh	Stevens 287 + 4 bis lots 10pp.	**CAMB** m.p.; **BMNHL** m.p. copy
May 23	ANON	Birds (21) Mam (6)	Engall, *Cheltenham* 130 (incl. 11 blank) lots 7pp.	DIP; **BMNHL** copy
June 18	HAWKINS (Rev. Herbert S.)	Birds Ool	Stevens	*Athenaeum*, June **1867**:775
June 19	ANON	Birds HH Br ins Nh	Stevens	*Athenaeum*, June **1867**:775
June 25-26	ANON	Sh	Stevens	*Athenaeum*, June **1867**:775
July n.d.	CLARK (Rev. Hamlet)	Entomological lib	Stevens	Allingham (1924:70)
July 16-17	(a) SMITH (Samuel), Liverpool (b) ANON	(a) Sh (b) Sh	Stevens	*Athenaeum*, July **1867**:3
November 15	CARTER (Samuel)	World lep World col	Stevens	*Athenaeum*, Oct. **1867**:483
December 13	OWEN (T.B. Bulkeley)†	Sh Cor Fos	Stevens 237 lots 10pp.	**NMW**
December 13	(a) [MANN] (b) ANON	(a) Fos Min Sh (b) Fos Min Irr	Sotheby 200 + 3 lots 11pp.	**BMNHM** f.p.
December 18	[PERKINS (W.), Brighton] †	Min (177) Fos (1)	Puttick 318 lots 13pp.	**BML** p.n.
1868				
n.d.	HEWARD (R.)	Herb Lib	Stevens	Allingham (1924:69)
February 21	HARRISON (H.W.)†	Sh	Stevens	*Athenaeum*, Feb. **1868**:195

[1] 'The American to whom a large part of Parkinson's collection was sold was G.W.Featherstonhaugh, F.G.S., whose museum was later destroyed by fire' (Dr. A.D. Morris *in litt.*). Cf. Parkinson sale iv: 1827.

1868–1869	SOURCE	CONTENTS	AUCTIONEER & SALE CAT.	REFERENCE
March 10	BRYSON (Alex.)† incl. Wm. NICOLL	Min	Stevens	*Athenaeum*, Feb. **1868**:270 Allingham (1924:55)
April 24	(a) CHANT (John)† (b) – (e) ANON	(a) Br lep (b) Lib (c) Exotic ins (d) Ool (e) Deer horns Antelope horns	Stevens 316 lots 12pp.	**BMNHE**
June 30	DESVIGNES (Thomas)	Br ins Entomological lib	Stevens	*Athenaeum*, June **1868**:847 Allingham (1924:152)
August 14	(a) CANTOR (Dr.), Calcutta (b) – (e) ANON	(a) Herb Birds Nh (b) Sh (c) Fos Indian min (d) Fos Sh Nh (e) Birds Nh	Stevens 296 + 4 bis lots 10pp.	**NMW**
October 27	ANON	Min (31) Fos (1) Herb (1)	Puttick 344 + 8 bis lots 15pp.	**BML** p.n. add
December 17	HAWKINS (Rev. Herbert)	Dup ool	Stevens 370 lots 15pp.	**BMNHZ**
1869 February 18	ANON	Sh	Stevens [178 + 8 bis lots] [11pp.]	**NMW** p. inc. a
March 5	(a) – (b) ANON	(a) Birds (b) Fos Birds Mam Ins Nh	Stevens 215 lots 8pp.	**BMNHZ**
April 27-28	(a) TROUGHTON (N.), Coventry† (b) ANON	(a) Br birds Br ool (b) For birds	Stevens 674 lots 20pp.	**BMNHZ**
May 3	(a) COOPER (Abraham) (b) HEARSEY (Brig.Gen. John)	(a) Br ins (b) Exotic lep	Stevens 278 lots 12pp.	**BMNHZ**; J s.p.s.n.
June 22	[ANON]	Sh	[Stevens] [231 lots] [14pp.]	CAMB m.p. in **BMNHL** m.p. inc. copy
June 25	HARTWRIGHT (J.H.)†	Br lep Br col Lib, 189	Stevens 297 lots 19pp.	J s.p.s.n.
July 28	(a) DUFF (R.) (b) – (c) ANON	(a) Ool (b) Ool (c) Birds	Stevens 239 lots 8pp.	**BMNHZ**
October 29	(a) HAMMOND (W.O.) (b) HOPLEY (E.)†	(a) Br lep (b) Br lep	Stevens 463 lots 27pp.	**BMNHE** m.p.; J p.m.n.; H
December 9	HARVEY (Rev. R.)	Ool	Stevens 310 lots 18pp.	**BMNHZ**

1869 – 1871	SOURCE	CONTENTS	AUCTIONEER & SALE CAT.	REFERENCE
December 10	(a) BREWER (J.A.) (b) BLACKMORE (T.) (c) ANON	(a) Br col (b) Lep (c) Br lep	Stevens 212 lots 12pp.	J p.m.n.
1870 March 17-18	(a) KNAGGS (H.G.) (b) BLACKBURN (Rev. Thomas) (c) ENGLEHEART (N.B.) (d) ANON	(a) Lep (b) Br col (c) Br ins For ins (d) Br ins For ins	Stevens 548 lots 26pp.	**BMNHZ; J** s.p. s.n.; **BMNHE** inc.
March 30-31	HAMILTON (William J.)†	Sh Min Fos Lib	Stevens 640 lots 32pp.	NMW
May 18	'An Indian officer' [=CRIPPS]	Sh Cab	Puttick 359 + 6 bis lots 12pp.	**BML** p.n.
July 12	(a) CONNEL (Prof., F.R.S.)† (b) PAUL (Mrs.) (c) ROSE (Prof.) & DEUCHARS (Prof.), both of Edinburgh (d) – (e) ANON	(a) – (c) Min (d) Birds Ool (e) Min (chiefly Cornish) Sh	Stevens 325 lots 16pp.	**BMNHM** m.p.
August 9	ANON	Birds Ool Ins Nh HH	Stevens 269 lots 8pp.	J s.p.s.n.
November 22	(a) FLEMING (Prof.)†[1] (b) ANON	(a) Sh Fos (b) Ins Nh	Chapman, *Edinburgh* 120 lots 7pp.	AR; **BMNHL** copy
November 22	(a) FENN (C. & J.) (b) EDMONDS (Abraham)† (c) EVANS (W.F.) (d) ANON	(a) – (b) Br lep (c) Br ins For ins (d) Lep	Stevens 307 lots 16pp.	**BMNHE** s.p.; J p.m.n.
1871 N.d.	DOUBLEDAY (Henry)	"Birds sold by auction in 1871"		Sherborn (n.d.); W.H.Mullens & H.K.Swann, *Bib.Br.Ornith.* (1916-17),p.175
N.d.	"A gentleman resident in South America for many years"	Min	Stevens	Allingham (1924:55)
N.d.	WATSON (John)	Exotic lep	Stevens	Allingham (1924:131), Horn & Kahle (1937:297)
February 14	(a) SCOTT (John) (b) – (e) ANON	(a) Br lep (b) Lep (c) Birds (d) Br macrolep (e) Col	Stevens 292 lots 15pp.	J s.p.s.n.; **BMNHE** inc.
March 20	ANON	Fos Min, 9	Puttick 397 + 3 bis lots 2 + 13pp.	**BML** p.n.

[1] John Fleming (1785-1857), author of *History of British Animals* and many other important binomial works – see *Roy.Soc.Catalogue of Scientific Papers.* Much of the material from this sale is now in the Royal Scottish Museum, Edinburgh (*teste* C.D.Waterston *in litt.* 21 ii 1973)

1871 – 1873	SOURCE	CONTENTS	AUCTIONEER & SALE CAT.	REFERENCE
August 17	ANON	Birds Fos HH, 18	Puttick 169 + 13 bis lots 2 + 8pp.	**BML** p.n.add.
October 19-20	(a) CHAUMETTE (A. de la) (b) SCOTT (John) (c) MERCER (A.H.) (d) – (e) ANON	(a) Br lep European lep (b) Br macrolep (c) Lep (d) Sh (e) Entomological lib	Stevens 658 lots 32pp.	**J** m.p.; **BMNHE** inc.
1872 February 15-16	HARPER (Dr.)	Br lep	Stevens 462 lots 27pp.	**J** p.m.n.; **BMNHE** inc.
April 9	WOMBWELL'S MENAGERIE	Quad Rept Birds	Buist, *Edinburgh* 186 lots	*The Scotsman*, 8.iv. 1872, pp.2b, 8f; *The Edinburgh Courant*, 10.iv. 1872, p.4a.
May 8-11, 13-16	PURNELL (B. Purnell)	15 v: Birds Mam Nh, 46	Sotheby 1456 + 2 bis lots 97pp.	**BML** p.n.
May 10	(a) – (c) ANON	(a) Min (b) Birds (c) Sh Fos	Stevens 304 lots 12pp.	**BMNHM** s.p.
May 28	(a) ANON (b) WALLACE (A.R.) (c) FORTUNE (R.) (d) NASH (J.), Canterbury (e) LAYCOCK, Sheffield (f) ANON	(a) For lep (b) Exotic lep (c) Chinese col Japanese col (d) Br lep (e) Br lep (f) Br lep	Stevens 290 lots 15pp.	**J** m.p.m.n.; **H**; **BMNHL** copy
June 6	HARTING (J.E.)	Ool Birds nests Birds	Stevens 260 lots 34pp.	**BMNHZ**
June 28-29	DOUBLEDAY (Henry)	Lib Irr	Stevens 441 lots 20pp.	**J** p.s.n.
October 25	(a) STANDISH (Joseph) (b) ANON	(a) Br lep (b) Lep Entomological lib	Stevens 315 lots 19pp.	**J** m.p.s.n.
October 29	ANON	Birds Quad Ins Fos Min Sh Skulls Nh (all these groups from India)	Stevens 389 lots 19pp.	**BMNHZ**
October 29 – November 1	BACON (Rev. Thomas), Kingsworthy Rectory, near Winchester	Sh Min Fos Nh, 15	Gadsden & Ellis, *house sale* 982 (incl. 69 blank) + 3 bis lots 64pp.	**VA**
November 13	GRAY (George Robert)†	Lib	Sotheby 201 lots 12pp.	**BMNHZ**
1873 N.d.	PERKINS (Algernon), Hanworth Park	Min	Stevens	Allingham (1924:56)

1873 – 1874	SOURCE	CONTENTS	AUCTIONEER & SALE CAT.	REFERENCE
n.d.	MARSHALL (William), Clayhill, Enfield.	Br ins Nh	Stevens	Horn & Kahle (1936:166)
March 14	(a) – (b) ANON	(a) – (b) Br lep	Stevens 257 lots 12pp.	J s.p.s.n.
March 27-28	RUCKER (S.)	Sh	Stevens 494 lots 20pp.	NMW
April 8	(a) WOODS (W.G.) (b) – (e) ANON (f) BROWN (Rev. J.L.)	(a) – (b) Br lep (c) Br hem (d) Br ins (e) For ins (f) Br ins	Stevens 304 lots 18pp.	J p.
May 21	(a) NORRIS (Thomas), Preston (b) ANON	(a) Br lep Lib (b) For lep For col	Stevens 355 lots 24pp.	BMNHZ; J p.s.n.
May 24	(a) ANON (b) [JERDON (Dr.)] (c) ANON	(a) Min (b) Japanese birds (c) Ins	Stevens 285 lots 11pp.	J p. of (c)
June 5-6	NORRIS (Thomas), Preston, Part I	Sh	Stevens 430 lots 20pp.	NMW s.p.; J m.p. s.n.; BMNHL copy, SPD copy
June 7	LEESON (Dr. Henry Beaumont), F.R.S., F.L.S., Bonchurch, Isle of Wight†	Min	Stevens 291 lots 16pp.	BMNHM p.add.
June 21	(a) PARRY (Thomas)† (b) ANON (c) [LEWIS (George)] (d) MIVART (St.George)	(a) Entomological lib Br lep Br col (b) Br lep (c) Dup Br col (d) Animal skeletons Animal skins	Stevens 242 lots 18pp.	J m.p.s.n.
July 22	'A well-known conchologist'	Sh ('chiefly from Mauritius and its dependencies')	Stevens 234 + 23 bis lots 20pp.	NMW
July 29-30	NORRIS (Thomas), Preston, Part II	Sh	Stevens 478 (431-908 lots) 23pp.	BMNHZ copy; SPD copy; NMW inc.
July 30	NORRIS (Thomas), Preston, Part III	Lib	Stevens 95 (909-1003 lots) 8pp.	J
1874 January 27-31, March 2-4	ASKEW (Henry William), Conishead Priory, Ulverstone, Lancashire	30 i: Min Fos Sh Birds Zoophytes Ins HH, 73	Burton, *house sale* 1132 + 5 lots 53pp.	VA
March 6	ANON	Ins Min Sh Fos Cor Nh, 12	Puttick 368 lots 11pp.	BML p.n.add.
April 17	(a) – (d) ANON	(a) Br lep (b) Exotic lep Exotic col (c) Swedish ool Norwegian ool (d) Turkish ool	Stevens 279 lots 12pp.	J s.p.

1874 – 1876	SOURCE	CONTENTS	AUCTIONEER & SALE CAT.	REFERENCE
April 18,20	ANON; WILSON (Wm.), F.R.S.†	(a) Sh Fos Min Nh (b) Bot lib	[Stevens] [598 + 11 bis lots] [26pp.]	NMW inc.
April 29-30, May 2	SAUNDERS (W. Wilson), Hillfield, Reigate.	Fos Sh Birds Fish Ins App Ool HH Cor Min Nh Cab	Lees, *house sale* 931 lots 40pp.	J; H
May 19-20	ANON	19 v: Min Sh Fos, 103	Puttick 354 + 11 bis lots 2 + 14pp.	BML p.n.
1875 January 29	YORK (Col. Philip James)	Lib	Stevens	*Athenaeum*, January, **1875**: 107
February 26	(a) – (d) ANON (e) BLACKMORE (T.) (f) – (h) ANON (i) EVANS (W.F.)†	(a) Br lep (b) Br col (c) Br lep (d) Exotic ins (e) For butts (f) Br lep (g) Ool (h) Lib (i) Lib	Stevens 410 lots 23pp.	J s.p.
April 23	(a) EDWARDS, Richmond (b) ANON (c) ANON	(a) Br lep (b) Br lep (c) Exotic butts.	Stevens 219 lots 14pp.	J s.p.s.n.
April 26	(a) REYNOLDS (F.A.) (b) YORKE (Colonel)†	(a) Min Sh Fos Cab (b) Min	Sotheby 287 lots 15pp.	BML p.n.; S p.n. copy
April 27	(a) WRIGHT (Bryce M.), Great Russell St. (b) ANON	(a) Min (b) Fos	Stevens 246 lots 16pp.	BMNHM m.p.
June 29	[BEWLEY]	Sh (incl. 'type shells figured by Reeve')	Stevens 290 lots 12pp.	NMW
August 31	ANON	Min	Stevens	*Athenaeum*, August **1875**: 135
September 29, October 1	DOUBLEDAY (Henry), High Street, Epping†	Lib Ins Birds Irr	McKenzie, *house sale* 603 lots 28pp.	J
November 23	(a) – (d) ANON (e) SAXBY (Dr.) (f) – (h) ANON	(a) Br lep (b) For ins (c) European lep (d) Ool (e) Ool (f) Sh (g) HH (h) S. African birds, S. African rept. S. African animals.	Stevens 368 lots 17pp.	J s.p. (of Ins)
1876 January 22,24	REYNE (Dr. P.), Marseilles	Min	Stevens	*Athenaeum*, January **1876**: 39
February 11	WATTS-RUSSELL (Jesse), Ilam Hall, Staffordshire	Min	Sotheby 170 lots 11pp.	BML p.n.; S p.n. copy

1876–1878	SOURCE	CONTENTS	AUCTIONEER & SALE CAT.	REFERENCE
May 30	CRAMER (Charles), Eastmount, Ryde, Isle of Wight	Min	Sotheby 256 lots 14pp.	BML p.n.; **S** p.n. copy
June 29–July 1	(a) ANON (b) SAUNDERS (W.W.), et al	(a) Col Lep Hym Hem Rept Ool (b) Sh	Sotheby 951 lots 43pp.	BML p.n.; **S** p.n. copy
November 28	[ANON]	Min	Stevens	MS. note by L.Fletcher in BMNH Min coll. Register (entry BM 50811) *per* P.Embrey
1877 N.d.	BAGOT (Colonel)	Mam	Stevens	Allingham (1924:63)
March 9-10, 12	BROWN (Edwin), Burton-on-Trent†	Br ins For ins Br herb Sh Birds	Stevens 915 lots 58pp.	**J** p.m.n.; **H**; SAR
March 14	ANON	Min, 4	Sotheby 158 lots 17pp.	BML p.n.; **S** p.n. copy
March 15-16	(a) ANON (b) MOORE, Stamford Hill (c) BLACKMORE (T.) (d) – (f) ANON (g) WALLIS (G.) (h) DIX (Rev. Joshua)	(a) Br ins (b) Br lep (c) For ins (d) For aculeate hym (e) Ins (f) Mauritian sh (g) S.American sh (h) For ferns	Stevens 514 lots 24pp.	**J** m.p.m.n.; **H**; SAR; **BMNHL** copy
March 19	MARSHAM (Hon. Robert)	Min Fos	Stevens	*Athenaeum*, March **1877**:307
May 7-8	(a) BELCHER (Admiral Sir Edward), K.C.B.† (b) – (c) ANON	(a) Sh (b) Sh (c) Fos Cor Indian ool Nh	Stevens 485 + 4 bis lots 20pp.	NMW
May 17	SAUNDERS (Howard)	Dup birds Dup birds' nests Dup ool	Stevens	*Athenaeum*, May **1877**:594; Allingham (1924:172)
June 18	(a) – (d) ANON (e) WALKER (Francis)† (f) IRVINE (Alexander) (g) ANON	(a) Birds Rept Fish HH Nh (b) Br ins (c) For ins (d) Br lep (e) Ins (f) Herb (g) Sh Nh	Stevens 286 lots 20pp.	**J** m.p.s.n.
July 9	(a) CULLEN (Dr.) (b) ANON	(a) Turkish birds Turkish ool (b) Sh Fos Min Nh	Stevens 357 + 10 bis lots 12pp.	NMW
July 12	FINLAY (George)†	Min, 1	Sotheby 2 lots 4pp.	BML p.n.; **S** p.n. copy
August 28-29	WHITFIELD (R.G.)†	HH Nh	Stevens 448 lots 20pp.	**J**
1878 N.d.	WARD (E.H.)	Birds	Stevens	Allingham (1924:172)

1878–1879	SOURCE	CONTENTS	AUCTIONEER & SALE CAT.	REFERENCE
February 25	ANON	"Insects from Trinidad", 15	Sotheby 230 + 5 bis lots 18pp.	BML p.n.
April 9-11	(a) MURRAY (Andrew) (b) LATHAM (A.G.), Manchester	(a) Ins (mostly col) Lib (b) Br lep For ins Exotic lep Br col Br ool	Stevens 795 lots 54pp.	CH m.p.m.n.; J p.m.n.
April 30	(a) WESTMINSTER AQUARIUM (b) ANON	(a) Australian nh (b) Several small collections of sh & ins	Stevens	*Athenaeum*, April 1878:463
May 24	(a) SMEE (General), Reigate† (b) – (c) ANON (d) HARPER (J.O.), Norwich	(a) Sh (b) Ins (c) Min Fos (d) Birds' skeletons Animal skeletons	Stevens 378 + 1 bis lots 14pp.	NMW; J m.p.s.n.
July 30	(a) – (d) ANON	(a) Min Fos (b) Sh (c) Birds (d) Ool	Stevens 323 lots 16pp.	J s.p.
September 6	ANON	Bot lib	Stevens 263 lots 12pp.	J m.p.f.n.
October 25	(a) BARLOW (F.), Cambridge (b) SPILLER (A.J.) (c) – (e) ANON	(a) Br lep (b) Br lep (c) Lep (d) Lib (e) Birds Min Nh	Stevens 330 lots 16pp.	J m.p.m.n.; CH
October 29	ANON, 11 Finchley Rd., St. John's Wood.	Ool, 63	Hodge *house sale* 291 lots 19pp.	J
December 16	(a) – (c) ANON	(a) For col (b) Ins (c) Lib	Stevens 324 lots 18pp.	J m.p.s.n.
1879 N.d.	GOLDIE	Birds	Stevens	Allingham (1924:172)
January 24	(a) ANON (b) [BEDELL] (c) WILKINSON† (d) HOEY (e) WHYTE & Co.	(a) Br lep (b) Br lep (c) Ins (d) Hym (e) Ceylon butts	Stevens 260 lots 16pp.	J m.p.f.n.; CH
June 6	NEVILLE (Lady Dorothy)	Lib Min Irr	Stevens 324 lots 18pp.	J m.p.m.n.
June 10	(a) – (f) ANON	(a) Min Fos Sh (b) Fos (c) Sh (d) Birds (e) Birds Mam Nh (f) Ool	Stevens 337 lots 12pp.	J m.p.m.n.
June 17-19	WARING (Samuel Long), The Oaks, Gipsy Hill, Upper Norwood†	18 vi: Birds Fos Min Sh Lep Nh, 48	Foster 605 + 3 lots 27pp.	EKW p.n.; VA p.n.
July 21	"A German Professor"	Min	Sotheby 237 lots 14pp.	BML p.n.; S p.n. copy
September 9	(a) – (d) ANON	(a) For lep (b) – (c) Br lep (d) Sh Fos Crustacea Nh	Stevens 260 lots 13pp.	J s.p.s.n.

1879 – 1880	SOURCE	CONTENTS	AUCTIONEER & SALE CAT.	REFERENCE
November 7	[BOUCARD, Natural History dealer]	Br birds For birds Ins Sh Nh	Stevens 309 lots 11pp.	J s.p.s.n.
1880				
n.d.	LOW (Dr.)	Exotic butts	Stevens	Horn & Kahle (1935:159)
n.d.	RUCKER (Sigismund)	Birds	Stevens	Allingham (1924:172)
May 6-7	(a) – (e) ANON	(a) Exotic butts (b) Br lep Br col Exotic lep Exotic col (c) Lib (d) Fos (e) Birds Ool Sh Min Nh	Stevens 687 lots 27pp.	CH; J s.p.f.n.
May 8	MURRAY (W. Cleghorn), 3 Clarendon St., Edinburgh	Egg of Great Auk Irr	Dowell, *Edinburgh*	Parkin (1911:10)
May 12-14	STEPHENSON (Robert)	12 v: Birds, 43	Norfolk, *Beverley, Yorks.* 870 (incl. 84 blank) + 9 bis lots 45pp. + 2 plates	VA
June 8	'A Gentleman many years resident in Amboyna'	Sh	Stevens 205 + 1 bis lots 10pp.	NMW m.p.f.n.
June 21-22	[TAYLOR (Thomas Lombe), Starston, Norfolk]	Sh	Stevens 559 lots 24pp.	NMW
July 2	(a) SMITH (F.)† (b) ANON (c) WATERHOUSE (G.R.) (d) ANON	(a) Lib Ins App (b) 2 Great Auk's eggs (c) Lib (part of) Fos Sh	Stevens 309 lots 20pp.	J p.f.n.
July 6	BARRY (Dr. Martin)	Birds Min Fos Sh Cab Mam Butts Col Fish Rept Nh, 236	Puttick 340 + 28 bis lots 2 + 14pp.	BMNHM f.p.
July 20	(a) – (e) ANON	(a) Birds (b) Ins (c) Fish (d) Fos Ool Sh Nh (e) Lib	Stevens 362 lots 18pp.	J m.p.
August 4	(a) BELL (Professor Thomas), F.R.S., F.Z.S., F.G.S., late Secretary of the Royal Society† (b) ANON	(a) Min Fos Sh Crustacea Br lep Exotic lep Col Ins (b) Min Fos	Sotheby 187 lots 10pp.	BML p.n.; S p.n. copy
August 13	(a) STANDISH (F.O.) (b) MEEK (E.G.) (c) WARING	(a) Br lep (b) Ins (c) Br col	Stevens 262 lots 16pp.	J m.p.f.n.
September 10	(a) ANON, Wimbledon (b) ANON (c) FELDER (Dr.), Vienna (d) – (e) ANON	(a) Exotic butts (b) Sh (c) European & N. Asiatic butts (d) Birds Exotic butts Nh (e) Ool	Stevens 432 lots 20pp.	J
October 1	ANON	Lib Irr	Stevens 232 lots 12pp.	J p.f.n.

1880 – 1882	SOURCE	CONTENTS	AUCTIONEER & SALE CAT.	REFERENCE
October 8	ANON	Exotic Lep	Stevens 286 lots 14pp.	J m.p.; CH; H
December 3	(a) KALTENBACH (J.H.) (b) – (e) ANON	(a) N.German ins (b) Br lep (c) Ins (d) Tortrices (Br) (e) Ins	Stevens 295 lots 14pp.	J m.p.
1881 January 14	(a) ANON (b) FARRE (Dr.) (c) ANON (d) JOHNSON (C.) (e) – (f) ANON	(a) Sh (b) Sh (c) Fish (d) Sh Fos Fish Min Herb Nh (e) Birds (f) Lib	Stevens 397 lots 18pp.	J m.p.m.n.
January 17-20	BELL (John)†, Part I	17 i: Min Fos Sh Ool Cor Birds Cab Recent crustacea, 8	Morrison, Dick M'Cullock, *Glasgow* 815 lots 60pp.	NPG; VA
February 11	(a) TURNER (J. Aspinall), Manchester (b) HUNTER (J.)	(a) Exotic col (b) Br lep	Stevens 273 lots 16pp.	J p.m.n.; H; **BMNHL** copy
May 4	GOULD (John), F.R.S.†	Ornithological lib	Puttick 301 lots 2 + 18pp.	**BML** p.n.add; **CAMB** p.n.
May 17	(a) – (c) ANON	(a) Min Fos Sh (b) Birds Ins Nh (c) Ins	Stevens 232 lots 10pp.	J f.p.
June 14	WESTON (W.P.)	Br lep Br col	Stevens 266 lots 14pp.	**BMNHE**; J p.m.
July 12-15, 18-20	DAUBENAY (George Matthews), Alstone Lodge, Cheltenham†	13 vii: Sh Min, 4	Engall Sanders, *house sale* 1582 (incl. 69 blank) + 3 bis lots 87pp.	NG
July 19	(a) BIRCHALL (Edwin) (b) – (c) ANON	(a) Br lep (b) Fos Sh (c) Birds Mam	Stevens 382 lots 18pp.	J p. of (a); H; **BMNHL** copy
October 25	(a) BRIDGER (b) BENSHER [Bensher's 2nd collection] (c) – (d) ANON	(a) Ool (b) Birds (c) Lib (d) Ins	Stevens 441 lots 24pp.	J s.p.s.n.
November 21-22	POPHAM (Francis Leyborne)	21 xi: Birds, 23	Christie 231 lots 13pp.	C p.n. add; VA
December 13	(a) – (c) ANON	(a) Sh (b) Birds Mam HH Nh (c) Birds	Stevens 336 lots 16pp.	J s.p.s.n.
1882 N.d.	LABREY (Beebee Bowman)	Butts	Stevens	Horn & Kahle (1935:146)
February 10-11	(a) – (d) ANON (e) DAVISON (W.) (f) – (g) ANON	(a) Lib (b) Sh Min Fos Ool Lep (c) Ins (d) Birds HH (e) Exotic birds (f) Rept Fish (g) Birds	Stevens 562 lots 24pp.	**CH** f.p.f.n; J f.p.

1882–1883	SOURCE	CONTENTS	AUCTIONEER & SALE CAT.	REFERENCE
March 24	(a) – (f) ANON	(a) Ool (b) Sh (c) Sh Fos (d) HH (e) Fos Min Ool Lep Nh (f) Exotic lep	Stevens 452 lots 20pp.	J p.m.n.
April 25	(a) HUNTLEY (Marquis of) (b) – (c) ANON	(a) Birds (b) HH (c) HH Birds	Stevens 237 lots 8pp.	J
April 26	(a) – (e) ANON	(a) Ool (b) Exotic lep (c) Min (d) Canadian birds (e) Birds	Stevens 518 lots 12pp.	J m.p.f.n.
May 20	(a) – (b) ANON	(a) Br lep (b) Ins	Stevens 249 lots 12pp.	J m.p.; H.
June 5, 7	SANG (John), Darlington	Macrolep Microlep Ins	Stevens 497 lots 24pp.	J m.p.m.n.; BMNHE s.p.
June 20	(a) – (d) ANON	(a) Br ins (b) Ool (c) Norfolk birds (d) Travancore land sh	Stevens 306 lots 14pp.	J m.p.
July 15	(a) SANG (John), Darlington (b) – (c) ANON	(a) Macrolep Microlep (b) Br lep European hym (c) Lib	Stevens 602 + 1 bis lots 31pp.	BMNHE f.p.
July 18-19, -22	(a) STUART (Mrs.), Bampton, Oxon† (b) ANON (c) HINDSON (Isaac), Kirkby Lonsdale, Westmorland (d) – (e) ANON (f) CRICHTON (A.W.)† (g) – (i) ANON	(a) Br lep Ool Sh (b) Lib (c) Fos Min Sh (d) Ool (e) Sh (f) Ool Birds Birds' nests (g) Br lep (h) Br lep (i) Fos Min Nh (j) Sh Birds	Stevens 1156 lots 46pp.	J m.p.; NMW (1st day only)
November 10	(a) PREST (Wm.), York (b) – (c) ANON	(a) Br macrolep Br Microlep (b) Lep Col (c) Birds Mam	Stevens 343 lots 18pp.	CH s.p.; J p.m.n.
December 8	(a) – (d) ANON	(a) Birds Ool (b) Lib (c) Exotic lep (d) Herb	Stevens 325 lots 14pp.	J f.p.
1883 March 9	(a) PINDER (Dr.) (b) – (c) ANON	(a) Fos (b) Lep (c) Br lep	Stevens 358 lots 16pp.	J m.p.m.n.
March 18	FOOTIT (W.F.), Croydon†, Part I	Ool	Stevens	*Athenaeum*, May 1883:590
March 28	FOOTIT (W.F.), Croydon†, Part II	Birds	Stevens	*Athenaeum*, May 1883:623, 654
June 1	FOOTIT (W.F.), Croydon†, Part III	Birds Ool	Stevens	*Athenaeum*, May 1883:655
June 11-12	(a) – (b) ANON (c) BAYLEE (Rev. J.), Stroud (d) – (f) ANON	(a) Br lep (b) Exotic lep Exotic col (c) Fiji nh (d) Sh Min Ool Birds Ins Nh Lib (e) Br lep (f) For lep For col	Stevens 679 + 7 bis lots 28pp.	CH

1883 – 1884	SOURCE	CONTENTS	AUCTIONEER & SALE CAT.	REFERENCE
June 19	(a) COOKE (Benjamin), Southport† (b) – (c) ANON (d) RUSKIN (John) (e) WARD (Dr. Ogier), Eastbourne	(a) Br ins (b) Lib (c) Sh (d) Min (e) Fos	Stevens 423 lots 24pp.	J s.p.s.n.
June 26	FOOTIT (W.F.), Croydon†, Part IV	Birds Ool	Stevens	*Athenaeum*, June 1883:751
June 29	(a) FOOTIT (W.F.), Croydon† Part V (b) – (c) ANON	(a) Br lep (b) Birds (c) Min Lib	Stevens 302 lots 12pp.	J m.p.
July 31	BOUCARD (Adolphe), Part I	Lib	Stevens 255 lots 18pp.	J p.
[September] N.d.	HARRIS (Mrs. Sabina), Mount House, Newport, Isle of Wight†	Sh Min Fos Nh	Pittis, *house sale*	*Nature*, August 1883:cxxxix
October 9	(a) [HUMBLOT (L.)] (b) – (c) ANON	(a) Madagascar ins Madagascar nh (b) HH (c) Lib	Stevens 235 lots 10pp.	J m.p.m.n.
October 22-23	GRAY (John), Claygate, Esher & formerly Bolton	European col Lib	Stevens 604 lots 42pp.	J m.p.m.n.
1884 January 11	(a) SHEPPARD (E.R.), Lewisham, Kent (b) ANON	(a) Br macrolep Br microlep (b) For lep	Stevens 327 lots 16pp.	**CH;** J m.p.f.n.
January 14	(a) GLOYNE (C.P.) (b) ANON	(a) Sh (b) Min Fos	Stevens 299 lots 16pp.	**NMW** m.p.f.n.; J p.m.n.
March 17	(a) GLOYNE (C.P.) (b) ANON	(a) Min Ins (b) Lib	Stevens 303 lots 14pp.	J m.p.f.n.
March 20-21	HARPER (Philip H.), F.R.C.S.†, Part I	Br lep	Stevens 615 lots 28pp.	**CH; BMNHE;** J p.m.n.
March 24	(a) – (e) ANON. [F.S. BIRCHAM's ool coll. apparently incl. in this sale]	(a) Madagascar col (b) Sh (c) Ool (d) Ool (e) Min Fos Nh	Stevens 428 lots 16pp.	J f.p.
April 8	(a) – (c) ANON	(a) Min (b) Min Fos Lep Nh (c) Birds Br lep Nh	Stevens 234 lots 10pp.	**NMW** inc. (lacks pp.5-6, lots 55-127)
April 25	HANBURY (Robert), Ware	Lib	Stevens 162 lots 14pp.	J p.f.n.
May 8-9	HARPER (Philip H.), F.R.C.S.†, Part II	Br macrolep Br microlep Br col Lib	Stevens 1135 lots 36pp.	**BMNHE; CH;** J p.m.n.
May 12	CRISP (Dr.), Beaufort St., Chelsea	Birds Mam Lib	Stevens 329 lots 10pp.	J p.m.n.

1884 – 1885	SOURCE	CONTENTS	AUCTIONEER & SALE CAT.	REFERENCE
y 14	(a) LANG (Henry Charles) (b) WAILES (George) (c) ANON (d) HANSON (Samuel)	(a) Br moths (b) Br ins (c) Br lep (d) Ins	Stevens 246 lots 18pp.	J m.p.m.n.
y 19	(a) SHEARWOOD (G.P.) (b) – (c) ANON	(a) Exotic butts (b) Col (c) Lib Birds Sh	Stevens 270 lots 14pp.	BMNHE; J p.m.n.
y 23	(a) STEPHENSON (E.)† (b) 'A YORKSHIRE MUSEUM'	(a) Min (b) Min	Stevens	*Athenaeum*, May, 1884:619
y 29-30	(a) HOARE (Captain Edward) (b) MILLIGAN (Joseph)†	29 v: Min Fos, 49	Sotheby 418 + 2 bis lots 2 + 20pp.	BML p.n. add.
e 23	BIDWELL (Edward)	Ool	Stevens 262 lots 16pp.	J p.m.n.
y 7-9	[HARFORD (Th.)]	Sh Cab Lib	Stevens 684 lots 38pp.	NMW; J f.p.; SPD
y 15	(a) PREST (William), York† (b) SAUNDERS (Sir Sidney Smith) (c) – (d) ANON	(a) Br lep (b) European col European orth Br hym For hym (c) Lib (d) Ins	Stevens 340 lots 22pp.	J p.m.n.
y 29	(a) BIDWELL (Edward) (b) – (d) ANON	(a) Ool (b) Ool (c) Min Fos Nh (d) Lib	Stevens 313 lots 16pp.	J m.p.m.n.
tober 14	(a) – (e) ANON	(a) Lep Col (b) Min Sh Fos (c) Herb HH (d) Shetland & other lep (e) Birds Ool HH	Stevens 286 lots 12pp.	J m.p.m.n.
85 uary 23	RUSSELL (William), Onslow Gardens†	Sh Fos Min Birds Ool Mam Ins	Stevens 366 lots 14pp.	J m.p.m.n.
uary 26-30, oruary 2-6, 2	SILVER (Rev. F.), Norton, near Market Drayton	Ool Min Fos Sh Birds Lep Nh, 138	Edwards, *house sale* 2670 + 1 bis lots 111pp.	VA
oruary 6	TERRY (William), Fulham†	Min	Stevens 214 lots 12pp.	J s.p.s.n.
oruary 7	TERRY (William), Fulham†	Lib	Stevens 290 lots 24pp.	CH f.p.; J m.p. m.n.
rch 9	GUTHKUNST (H.G.), Stuttgart	Sh	Stevens	*Athenaeum*, March 1885:299
rch 30-31	RAINE (F.), Durham	Birds Animals	Stevens 594 lots 26pp.	J f.p.f.n.
ril 13	(a) HENNAH (Rev. W.), East Cowes, Isle of Wight† (b) – (e) ANON	(a) Sh (b) Min Fos (c) Br lep For lep (d) Sh (e) Birds	Stevens 456 lots 16pp.	NMW; J

1885–1886	SOURCE	CONTENTS	AUCTIONEER & SALE CAT.	REFERENCE
May 16	PARRY (Frederick John Sidney), 18 Onslow Square, Brompton†	Col Entomological lib	Stevens 327 lots 22pp.	J p.n.; BMNHE
May 19	ANON	Lib Br lep Birds HH Sh Br ool American birds American ool Br Guiana birds	Stevens 364 lots 18pp.	J
May 21-23	KNIGHTON (Sir William W.), Blendworth Lodge, Hampshire†	21 v: Min, 1	Christie 534 + 7 bis lots 44pp.	C p.; BMP; VA
May 29	ANON	Sh	Stevens 271 + 16 bis lots 18pp.	NMW; J
June 9	ANON (several collections)	Min Sh Fos Ins Lib Nh	Stevens 376 lots 18pp.	J
July 8-11	SALPERTON PARK near Cheltenham	10 vii: Birds Quad, 63	Villar, house sale 1010 (incl. 49 blank) lots 36pp.	NG
August 7	(a) FIELD (Frederick)† (b) JONES (E.W.)†	(a) Min (b) Min	Stevens	Athenaeum, July 1885:99
October 9	(a) ANON (incl. much from coll. F.J.S.PARRY) (b) – (d) ANON	(a) Br col For col (b) Ool HH (c) Sh Min (d) Lep Hem Col	Stevens 321 lots 20pp.	J p.s.n.
November 6	LAW (J.S.), Southgate†	Lib	Stevens 333 lots 36pp.	J p.m.n.
November 10	(a) D'EMMICH (Gustave), Budapest (b) MURRAY (Rev. R.P.) (c) FOLLIOTT (George), Chester†	(a) European lep (b) For lep (c) Min	Stevens 299 lots 24pp.	J m.p.m.n.
December 14	(a) WHITE (Samuel), South Australia (b) ANON	(a) Birds (b) Exotic butts Ool Birds Min Rept Mam Sponges Cor	Stevens	Athenaeum, Dec. 1885:750
1886 January 22	BROWN (George Dransfield)†	Sh Cor Nh Lib	Stevens	Athenaeum, Jan. 1886:3
January 25	(a) ATKINSON (W.S.)† (b) ANON (c) PARRY, Canterbury (d) LEWIS (G.) (e) – (f) ANON	(a) Br lep (b) Indian lep (c) Br lep taken in 1885 (d) Br col (e) Bogota ins (f) Birds Ool	Stevens 368 lots 22pp.	J m.p.m.n.
January 27	(a) – (c) ANON	(a) Birds Ool (b) Br lep (c) Herb	Stevens 236 lots 10pp.	J s.p.s.n.
March 15	(a) MEEK (E.G.) (b) – (c) ANON	(a) Br lep (surplus stock) (b) Lep (Tortricidae) (c) Lib	Stevens 323 lots 18pp.	CH; J s.p.s.n.
April 12-13	(a) GILL (Battershell), M.D., F.R.C.S., Regent's Park (b) ANON	(a) Br macrolep Br microlep (b) Br lep For lep Exotic col.	Stevens 595 lots 38pp.	J p.n.; CH; BMNHE

1886–1887	SOURCE	CONTENTS	AUCTIONEER & SALE CAT.	REFERENCE
ril 19	(a) GLOYNE (C.P.) (b) – (c) ANON (d) CLARK-KENNEDY (Capt. Alexander W.M.) (e) ANON	(a) Min (b) Min (c) Lep Col (d) Birds For butts For ins Birds Nh (e) Br Ool	Stevens 462 lots 22pp.	J m.p.
y 11-14	WALKER (Edward Joshua), Shotley House	11 v: Birds HH, 8	Davison, *Newcastle upon Tyne* 914 lots 52pp.	VA
y 18	(a) CHOLMONDELEY (R.), Shrewsbury (b) – (d) ANON (e) CLARK-KENNEDY (Capt. A.W.M.)	(a) Birds (b) Br ins For ins (c) Br herb (d) Sh (e) Birds	Stevens 282 lots 14pp.	J m.p.m.n.
ne 17	JARDINE (Sir William)†, F.R.S., F.L.S.	Birds	Puttick 350 + 6 bis lots 32pp.	CAMB p.n.; J
ne 22-23	(a) – (b) ANON (c) ENGLE-HEART, Bedfont (d) – (i) ANON	(a) Sh (b) Lib (c) Fos Lib (d) Exotic ins (e) Exotic col (f) Bogota lep (g) Birds Mam (h) Sh (i) Min	Stevens 653 lots 36pp.	J m.p.m.n.; NMW s.p.
gust 10	(a) – (c) ANON	(a) Exotic lep (b) Br lep (c) Min Sh Fos Birds Nh	Stevens 333 lots 22pp.	CH; J m.p.m.n.
ptember 14	(a) – (b) ANON (c) [VERKREUTZEN] (d) – (e) ANON	(a) Bogota butts (b) Br lep Nh (c) Sh (d) Lep Nh (e) Min	Stevens 317 lots 26pp.	J s.p.s.n.
vember 23	(a) STOWELL (Rev. H.), Breadsell Rectory near Derby (b) ANON	(a) Br macrolep Br microlep (b) Exotic lep Cab App	Stevens 423 lots 24pp.	J m.p.m.n.; BMNHE; CH; H
cember 13	ANON	Mam HH New Britain birds Br lep Fos Natal lep	Stevens 245 lots 12pp.	J m.p.m.n.
87 d.	DIGBY (C.R.)	Br lep	Stevens	Horn & Kahle (1935:57)
d.	LINDEN (A.)	New Guinea col	Stevens	Horn & Kahle (1935:156)
uary 21	(a) MORLEY (W.A.) (b) MOSSE (G.S.) (c) LIVINGSTONE (C.) (d) PARRY	(a) Br lep (b) Br lep (c) Br lep (d) Lep	Stevens 305 lots 22pp.	J p.n.; SAR
bruary 7	INDIAN AND COLONIAL EXHIBITION	Bot Mam HH Nh, 42	Stevens 370 lots 18pp.	J m.p.s.n.
bruary 14	(a) ANON (b) LINDEN (Auguste) (c) – (d) ANON	(a) Birds (b) New Guinea birds (c) Aru Is ins Aru Is nh (d) Natal lep	Stevens 306 lots 14pp.	J m.p.m.n.
rch 28	(a) KUPER (Rev. C.A.F.), Trelleck, Monmouth (b) [GIBB] (c) ANON (d) [ANDERSON (E.)] (e) [MEEK]	(a) European col Exotic col Lib (lots 1-212) (b) Lib (lots 213-216) (c) Ins (lots 217-269) (d) Ins Br macrolep Br microlep (lots 270-280) (e) Lib (lots 286-346)	Stevens 346 lots 28pp.	J m.p.m.n.; H

1887–1888	SOURCE	CONTENTS	AUCTIONEER & SALE CAT.	REFERENCE
April 25	(a) ANON (b) BARTON [? BARLOW] (Captain), Bideford (c) – (e) ANON	(a) Min (b) Ool (c) Br lep Exotic lep Ins (d) S.African ins (e) Exotic birds Exotic mam	Stevens 336 lots 18pp.	J s.p.s.n.
May 9	(a) SANG (J.), Darlington† (b) – (g) ANON	(a) Lib Entomological app. (b) Lib (c) Indian mam (d) Ins (e) Br lep (f) Exotic lep (g) Mauritian sh	Stevens 342 lots 18pp.	NMW; J p.s.n.
June 13-14	(a) MITFORD (Robert H.), Hampstead (b) ANON (c) PACKMAN† (d) WALLACE (Dr.), Colchester (e) – (h) ANON	(a) Br lep (b) Birds (c) Br lep (d) Br lep (e) Lib (f) Br col (g) African ins (h) Min Sh Birds Exotic col	Stevens 626 lots 41pp.	J m.p.m.n.
July 11-12	(a) FRASER (J.M.)† (b) – (c) ANON	(a) Lep Col (b) Sh (c) Ool Min Birds Lep	Stevens 505 lots 26pp.	J s.p.f.n.
October 11	(a) – (d) ANON	(a) Ool Birds nests (b) Exotic lep Exotic col (c) Exotic butts (d) Sh Min Fos	Stevens 312 lots 18pp.	J f.p.f.n.
October 15	WOODCOCK (Dr.), R.N., The Museum, St. Andrews Road, Anstruther†	Fos Mam HH Fish Rept Sh Ins Min Birds Nh, 57	Chapman, *Edinburgh* 253 lots 12pp.	BML add. (letters and obituary from *East Fife Record* of 16 x 1885)
December 13	(a) – (c) ANON (d) BLACKALL (Walter), Folkestone (e) ANON	(a) S.African hh (b) Great Auk egg (c) Min (d) Lep (e) Br lep	Stevens 385 lots 25pp.	CH s.p.s.n.; J m.p.m.n.
1888 February 13	(a) – (e) ANON	(a) Min Fos (b) Sh (c) Nh (d) *Aepyornis* egg (e) HH	Stevens 352 lots 18pp.	J m.p.f.n.
February 20-23	ANON	20 ii: Min Sh, 15	Sotheby 750 lots 37pp.	BML p.n.; S p.n. copy
March 12	(a) WARREN (W.), Cambridge (b) – (d) ANON	(a) Br macrolep Br microlep (b) Lep (c) Col (d) Br mosses	Stevens 305 lots 16pp.	J m.p.m.n.; H; BMNHE p.of (a
March 12	(a) ANON (b) CHAMBERS (C.) (c) WISE (Mrs.) (d) – (e) ANON	(a) Sh HH (b) Ool (c) Birds (d) Great Auk egg (e) Sh	Stevens 328 lots 16pp.	J
March 22-23	WRIGHT (Bryce), late of 204 Regent St., bankrupt	Min (his stock of)	Stevens	*Athenaeum*, March 1888:29
April 5-6	[ZOUCHE (Lady)]	6 iv: Birds, 1	Christie 277 lots 16pp.	C n; VA; BML
April 16	(a) BACKHOUSE (F.W.) (b) ROOKE (W.D.F.)† (c) – (d) ANON	(a) Ool Birds Br lep HH Min Nh (b) Ool Min Nh (c) For lep (d) Br col	Stevens 371 lots 24pp.	J m.p.m.n.
April 23	(a) – (c) ANON	(a) Fos Col Nh (b) Br col Nh (c) Sh	Stevens 415 lots 22pp.	J f.p.f.n.

1888 – 1889	SOURCE	CONTENTS	AUCTIONEER & SALE CAT.	REFERENCE
ay 14	(a) – (e) ANON (f) SMITH (Rev. U.), Stony Stratford (g) – (i) ANON	(a) For lep (b) Lep (c) HH (d) Lib (e) Sh (f) Min Fos (g) Birds (h) Crustacea (i) Br birds	Stevens 355 + 21 bis lots 20pp.	J m.p.m.n.; NMW
ne 25-26	(a) SHERWILL (J.L.) (b) – (e) ANON (f) BIRD (J.) (g) – (i) ANON (j) SMITH (Rev. U.), Stony Stratford	(a) Assam butts (b) Exotic butts Lep Col (d) Br lep (e) Ool (f) Ool (g) Ool (h) Lib (i) Sh (j) Fos Min	Stevens 776 lots 47pp.	CH s.p.s.n.; J m.p.m.n.; H
ly 16-17	(a) KENWARD (Jas.) (b) – (d) ANON	(a) Br lep (b) Lep (c) Lib (d) Sh Min Fos	Stevens 397 lots 22pp.	J s.p.s.n.; H
ly 17	(a) BLADEN (W.Wells), Stone, Staffs. (b) BIRD (J.) (c) BARACÉ (Comte de)	(a) Ool (b) Ool (remainder of coll.) (c) Ool	Stevens 299 lots 20pp.	J p.m.n.
ugust 21	(a) ANON (b) SALL (Dr. J.), New York (c) – (d) ANON	(a) Ool (b) Ool (c) Col (d) Min Fos	Stevens 306 lots 18pp.	J s.p.s.n.
eptember 11	(a) [WATKINS (W.)?] (b) ANON	(a) Exotic lep (b) Col Br lep	Stevens 208 lots 16pp.	J m.p.m.n.; H
ctober 15	(a) EEDLE (T.), Canning Town (b) – (d) ANON	(a) Br lep (b) Larvae of lep (c) Exotic lep (d) Birds Ool	Stevens 404 lots 22pp.	J m.p.s.n.; H inc.
ovember 19	(a) – (b) ANON (c) LEWIS, Kensington (d) ANON	(a) Sh (b) Br lep Exotic lep (c) Col (d) Australian birds Australian mam	Stevens 321 lots 17pp.	J m.p.s.n; NMW; H
ecember 3	(a) – (g) ANON	(a) Min Fos (b) Br lep (c) Exotic butts (d) Ool (e) Birds (f) Lep (g) Sh	Stevens 367 lots 22pp.	J f.n.; H
ecember 10	(a) SHERWILL (J.L.) (b) WALLACE & BATES	(a) Indian lep (b) East Indies lep Amazon lep	Stevens 317 lots 22pp.	J p.m.n.; CH; H
889 ebruary 18	(a) MATHEW (G.F.) (b) – (g) ANON	(a) Dup exotic butts (b) Min (c) Min (d) Exotic lep (e) Lib (f) Ool (g) Mauritian sh	Stevens 421 lots 26pp.	J m.p.s.n.; H; CH f.p.f.n.
arch 25-26	(a) SHEPPARD (A.F.)† and SHEPPARD (Major E.)† (b) – (c) ANON	(a) Br macrolep Br microlep (b) Sh Exotic ins (c) Borneo lep Nias lep Nias col	Stevens 658 lots 39pp.	J p.n.; BMNHE s.p.
pril 15	(a) MATHEW (G.F.) (b) ANON	(a) Exotic lep (b) Br lep Exotic lep Exotic col.	Stevens 381 lots 28pp.	J m.p.m.n.; H
pril 29-30	(a) – (b) ANON (c) Collected by a Member of the CHALLENGER EXPEDITION	(a) Lib (b) Sh Min Ins Nh (c) Birds	Stevens 666 lots 33pp.	J m.p.s.n.; H
ay 13	(a) VINGOE, Penzance (b) BABINGTON (Prof. Churchill) (c) ANON	(a) Br birds (b) Sh Ool Cab Nh (c) Lep Fos Nh	Stevens 480 lots 28pp.	J f.p.f.n.; N

1889–1890	SOURCE	CONTENTS	AUCTIONEER & SALE CAT.	REFERENCE
May 20	(a) WOODFORD (C.M.) (b) ANON	(a) Exotic butts (b) Exotic butts	Stevens 261 lots 14pp.	**J** m.p.m.n.; C f.p.; H inc.
May 27	(a) SHELLEY (Capt. G.E.) (b) VINGOE, Penzance (c) ANON	(a) African butts Exotic lep Exotic col Br lep (b) Br birds (c) Lib	Stevens 388 lots 28pp.	**J** m.p.m.n.; N
June 17	(a) EVANS (Dr.), Bradford (b) – (d) ANON	(a) Sh (b) Br lep (c) Exotic lep Exotic col (d) Birds	Stevens 308 lots 18pp.	**J** s.p.f.n.; NM H
July 8	COLLINS (H.), Aldsworth House, Hants†	Br birds	Stevens 360 lots 24pp.	**J**
July 9	(a) – (b) ANON	(a) Sh (b) Br lep	Stevens 348 lots 17pp.	**J**
October 8	(a) – (d) ANON	(a) Br lep (b) Br lep (c) Exotic butts (d) Birds	Stevens 309 lots 17pp.	**J** m.p.m.n.
October 22	(a) LITTLE (Burgess)† (b) – (c) ANON	(a) Min Sh (b) Sh S.African butts (c) Sh Min	Stevens 357 lots 20pp.	**J**
October 29	(a) – (c) ANON	(a) Indian lep (b) Exotic lep (c) Exotic col	Stevens 305 lots 19pp.	**J** p.m.n.; H
November 12	(a) ANON (b) DAVIS (Morgan), Swansea (c) RYLANDS (P.H.), Manchester†	(a) Birds Ool (b) Lep (c) Lep	Stevens 318 lots 16pp.	**J** p.m.n.; H
November 20-22	HANSON (Samuel)†	22 xi: Br lep Br col Ins, 42	Puttick 607 lots 42pp.	**BMP**; **J** p.n.
November 26	(a) WOODFORD (C.M.) (b) – (d) ANON	(a) Solomon Isles butts (b) Exotic lep (c) European lep (d) Br lep	Stevens 321 lots 18pp.	**J** p.m.n.; H
December 10	ANON	Min Mam Birds Ool HH Lep	Stevens 378 lots 22pp.	**J** s.p.s.n.
1890				
January 9	ANON	Min (from Co.Cork, Ireland), 61	Sotheby 194 lots 11pp.	**BML** p.n.; S p. copy
January 27	(a) GORE† (b) – (c) ANON (d) [LAMBERT (C.J.)] (e) – (g) ANON	(a) Br col (b) Japanese lep (c) Solomon Is. butts (d) Exotic lep (e) Jamaican ins (f) Sikkim lep (g) Br lep	Stevens 396 lots 27pp.	**CH**; **J** p.m.n.

1890	SOURCE	CONTENTS	AUCTIONEER & SALE CAT.	REFERENCE
March 5-6	STAFFORD (W.), naturalist, Godalming, Surrey†	Br birds Mam Sh Ool Fos Br lep Br col For lep For col	Mellersh, *house sale* 592 lots 21pp.	BMNHZ
March 10	PERCY (Dr. John)†	Min	Stevens 381 lots 24pp.	J
March 24	(a) – (e) ANON	(a) Br macrolep Br microlep (b) Chinese butts (c) – (e) Exotic lep	Stevens 401 lots 24pp.	CH; J m.p.m.n.
March 25	(a) WALLER (Dr. Arthur), Gibson Street, Islington (b) – (e) ANON	(a) Br lep Sh Min Nh (b) Br birds (c) Min Fos (d) Exotic lep (e) Br birds	Stevens 339 lots 20pp.	J f.p.f.n.
March 31	(a) HARRIS (Rev. G.P.), Richmond, Yorks (b) – (c) ANON	(a) Br lep (b) Birds Ool (c) Exotic lep	Stevens 434 lots 25pp.	J p.f.n.; CH; N
April 14	WALLER (Dr. Arthur), Gibson Street, Islington	Lib	Stevens 267 lots 26pp.	J p.m.n.
April 22-23	VAUGHAN (Howard)†, Part I	Br lep	Stevens 501 lots 36pp.	CH p.m.n.; LC p; J; BMNHE
May 5	(a) WOOD (Joseph)† (b) ANSTRUTHER (Captain)† (c) ANON	(a) Lep Col Sh Fos Min (b) Birds HH (c) Br lep	Stevens 424 lots 22pp.	J m.p.s.n.
May 20	(a) VAUGHAN (Howard)†, Part II (final) (b) SCHOOLING (H.C.) (c) SYME (J. Boswell)† (d) BLISS†	(a) Br microlep (b) Br lep (c) Br lep Br col (d) Br lep	Stevens 336 lots 24pp.	CH p.m.n.; BMNHE; J p.m.n LC
May 22	(a) WHITAKER (J.), Ranworth Lodge, Notts (b) PERCY (Dr.)† (c) ANON	(a) Br birds (ex coll. F.Bond) (b) Col Lep Hym Hem Dipt (c) Australian birds	Stevens 349 lots 32pp.	J p.m.n; CH
June 2	(a) BLABER (James)† (b) PERCY (Dr.)† (c) MARSHALL (W.)†	(a) Fos (b) Sh Min Nh (c) Herb	Stevens 347 lots 18pp.	J
June 5	(a) BLISS† (b) ANON	(a) Br lep (b) Exotic lep	Stevens 244 lots 12pp.	J m.p.m.n.
June 9	ANON	Br ool Indian ool Indian birds	Stevens 410 lots 24pp.	J m.p.
June 16	BABINGTON (Dr.)†	Min Fos	Stevens 225 lots 14pp.	J
June 17	PERCY (Dr.), Gloucester Crescent, Hyde Park†	Sh	Stevens 232 lots 15pp.	NMW; J p.n.
June 24	SMITH (Cecil), Lydeard, Taunton	Br birds	Stevens 294 lots 20pp.	J p.n.

1890–1891	SOURCE	CONTENTS	AUCTIONEER & SALE CAT.	REFERENC
June 25	(a) – (c) ANON (d) MARSHALL (W.) (e) – (g) ANON	(a) Indian lep (b) Exotic lep Exotic col (c) Birds (d) Herb (e) Herb (f) Fos (g) Sh	Stevens 327 lots 20pp.	J s.p.f.n.
October 14	(a) – (e) ANON	(a) Exotic lep (b) Exotic ool (c) Br lep (d) Birds Mam HH (e) Lib	Stevens 422 lots 22pp.	J s.p.
November 25	(a) GOODMAN (Nevil), Cambridge† (b) BALY (Dr.), Warwick (c) – (e) ANON	(a) Br ins European ins (b) Col Hym (c) Exotic lep (d) Larvae of lep (e) Ool	Stevens 373 lots 25pp.	J p.m.n.; CH
November 26	(a) – (c) ANON	(a) Min (b) Min Fos (c) Birds Mam	Stevens 232 lots 13pp.	J
1891 N.d.	LIVINGSTONE (C.)	Br lep	Stevens	Horn & Kahle (1935:157)
January 26	(a) – (f) ANON [DONISTHORPE; GROSE-SMITH; Dr. DUDGEON; Mrs. BEVERIDGE; A. REIS; et al.]	(a) Exotic lep (b) Exotic col (c) Col (Cetoniidae) (d) Exotic ins (e) Indian lep (f) Exotic lep.	Stevens 431 lots 29pp.	J m.p.m.n.
February 23	(a) – (h) ANON	(a) Lep (b) Turkestan lep (c) Lep (d) Exotic lep (e) HH (f) Lib (g) Birds Ool (h) Sh	Stevens 406 lots 24pp.	J m.p.m.n.
April 6	(a) STRONG (C.E.)† (b) [? STRONG (C.E.)] (c) ANON	(a) Sh Cor Nh (b) Lib (mostly conchological) (c) Sh	Stevens 292 lots 24pp.	J s.p.s.n.
April 28-29	(a) CRAVEN (A.E.) (b) – (e) ANON (f) BENNETT (Wm.)† (g) – (k) ANON	(a) Sh (b) Sh (c) Sh (d) Exotic nh (e) Butts from Thibet & W. China (f) Br ins (g) Exotic lep (h) Birds Ool (i) Fos Min (j) HH Mam (k) Lib	Stevens 711 lots 39pp.	J m.p.m.n.; N
May 4	(a) EDKINS (W.) (b) MITCHELL (F.S.) (c) FIELD (Leopold) (d) ANON	(a) Ool (incl. *Aepyornis maximus*) (b) Ool Birds Nests (c) Ool (dup) (d) Birds Ool	Stevens 459 lots 43pp.	N
May 26	(a) – (j) ANON	(a) – (c) Sh (d) Anderman Isles sh (e) Br lep (f) Br macrolep Br microlep Col (g) Lep (h) HH (i) Ool (j) Fos Min	Stevens 443 lots 27pp.	J m.p.m.n.; NMW
June 10-11	RIMINGTON (J.W.), Part I	Min	Stevens 492 lots 33pp.	J
June 15	RIMINGTON (J.W.), Part II	Min	Stevens	*Athenaeum*, May **1891**:683
June 23	(a) – (i) ANON	(a) Br macrolep Br microlep (c) Br lep Col (d) Lib (e) Indian lep (f) Br macrolep Br microlep (g) Exotic lep (h) Nh (i) Br birds	Stevens 444 lots 32pp.	BMNHE; J m.p. m.n.

1891–1892	SOURCE	CONTENTS	AUCTIONEER & SALE CAT.	REFERENCE
July 6-9	BARCLAY (Sir David W.)†	Sh Cab Conchological lib	Stevens 1178 + 20 bis lots 72pp.	**SPD** p.f.n.; **NMW** p. (incl. J.C.MELVILL's purchases), **BMNHZ** p.f.n.; **J** p.n. (of lib, f. sh lots)
June 14	(a) – (h) ANON	(a) Sh (b) Sh Min Ins (c) Exotic lep (d) Exotic col (e) Exotic lep (f) Min (g) HH Mam (h) Birds Ool	Stevens 412 lots 24pp.	**J** m.p.f.n.
August 11	(a) – (e) ANON	(a) Lep (b) Ins in amber (c) Birds (d) HH Mam (e) Lib	Stevens 255 lots 16pp.	**J** m.p.m.n.
October 27	(a) RAYNOR (Rev. G.H.) (b) – (d) ANON	(a) Br lep (b) Indian lep (c) Birds (d) Exotic lep	Stevens 401 lots 21pp.	**BMNHE**; **J** m.p.m.n.; **CH**; **SAR** inc.
November 23-24	(a) GRUT (Ferdinand)† (b) SHEARWOOD (G.P.)† (c) COX (G. Stanley)	(a) Exotic col Br col Br lep (b) Br lep Exotic lep (c) Min	Stevens 677 lots 57pp.	**BMNHE**; **J** m.p.m.n.; **CH**; **SAR**
November 26	GRUT (Ferdinand)†	Lib	Stevens 392 lots 39pp.	**BMNHE**; **J** p.m.n.
December 8	(a) GRUT (Ferdinand)† (b) – (h) ANON	(a) Exotic col (b) Exotic lep (c) Exotic lep (d) Sh (e) Sh (f) Sh (g) Fos Min Sh (h) Rept Min HH	Stevens 372 lots 24pp.	**NMW**; **J** f.p.f.n.; **SAR**
1892 February 15	(a) ANON (b) MEEK (E.G.) (c) – (e) ANON	(a) Brazil lep (b) Canary Is. butts (c) Exotic lep Col (d) Min Sh Birds Nh (e) Mauritian sh	Stevens 485 lots 29pp.	**BMNHE**; **NMW**; **J** m.p.m.n.
February 29	FORBES (James), Chertsey†	Mam Fish Br birds For birds	Stevens 246 lots 16pp.	**J** p.m.n.
March 21	(a) – (f) ANON	(a) Br lep Exotic lep (b) Exotic lep (c) Exotic butts (d) Min (e) Sh (f) Sh	Stevens 442 lots 25pp.	**J** f.p.f.n.; **NMW** inc. (lacks title & last p.) m.p. (incl. J.C. MELVILL's purchases)
March 28	(a) CAMERON (E.S.), Evie, Orkney (b) – (f) ANON	(a) Birds (b) Birds (c) Ool (d) Exotic lep (e) New Zealand birds & mam (f) Fos Sh	Stevens 336 lots 20pp.	**J**; **N**
April 25	(a) SALWEY (R.E.) (b) ANON	(a) Br lep (b) Exotic lep	Stevens 359 lots 21pp.	**BMNHE**; **CH J** m.p.m.n.
May 9	(a) IRBY (Col. L. Howard) (b) REID (Capt. Savile) (c) ANON	(a) – (b) Br birds For birds Br ool For ool Ornithological lib (c) Lib	Stevens 264 lots 27pp.	**J** p.m.n.; **CH** f.p.
May 16	(a) NAISH (Arthur)† (b) SAUNDERS (Howard) (c) ANON	(a) Br lep (b) Birds (c) Fos Min Sh Nh	Stevens 403 lots 20pp.	**BMNHE**; **J** m.p.s.n.

1892–1893	SOURCE	CONTENTS	AUCTIONEER & SALE CAT.	REFERENCE
May 23	(a) PEARSON (Halford)† (b) ROUSE (G.) (c) – (d) ANON (e) BURNELL (Edward)† (f) – (g) ANON	(a) Br ool (b) Br ool (c) Mam HH Birds (d) Fos Min (e) Br lep Br col (f) Lep (g) Ool (Cuckoos)	Stevens 376 lots 19pp.	J m.p.m.n.; N
June 14-17	HUDSON (W.T.), Bleak House, Nelson Road, Great Yarmouth†	17 vi: Birds Fish, 6	Aldred, *Great Yarmouth* 1130 (incl. 24 blank) lots 45pp.	CLD
June 28	(a) [SMITH (Sydney), Walmer, Kent] (b) – (h) ANON	(a) Br macrolep Br microlep (b) Exotic lep Col (c) Br col Hym Neur (d) Exotic butts (e) Br sh Exotic sh (f) Herb (g) Birds (h) Fos Min	Stevens 361 lots 20pp.	J m.p.m.n.; CH
August 9	(a) COOPER (E.)† (b) – (f) ANON	(a) Br lep (b) Br lep (c) Exotic lep (d) Exotic col (e) Birds Ool Sh Min (f) Lep (bred *Lycaena dispar* Haw.)	Stevens 455 lots 24pp.	J f.p.f.n.; CH; N
October 4	HILL (G.R.), Stoneleigh, Erdington near Birmingham	Lep Cab	Ludlow, Roberts & Weller, *Birmingham*	*Athenaeum,* Sept. **1892**:403
November 8	(a) [PRESTON (Rev. Darcy)] (b) – (c) ANON (d) SWINHOE (Colonel) (e) – (f) ANON	(a) Br col (b) Br macrolep Br microlep (c) Lep (d) Exotic butts (e) Br lep Br col (f) Lib	Stevens 228 lots 12pp.	J p.f.n.
November 22	[COOKE & SON]	Ool Sh (incl. 'many Gordon Cummings's') Br lep Rept Nh	Stevens 343 lots 12pp.	J f.p.f.n.; N f.p.f.n.
December 12	(a) STANDEN (R.S.) (b) [COOPER (B), Stoke Newington] (c) VAUGHAN (Howard)† (d) [MÖLLER] (e) [GROSE-SMITH (H.)] (f) – (g) ANON (h) [DURRANT (Hartley)]	(a) Br moths (b) Br lep (c) Br macrolep Br microlep Lib (d) Exotic lep (e) New Ireland lep (f) Ool (Cuckoos) (g) African hh (h) Macrolep from Melbourne, Victoria	Stevens 315 lots 20pp.	J m.p.m.n.
1893 February 13	(a) HAILES (H.F.), Secretary Quekett Microscopical Club† (b) REEVES (W.W.), former Asst. Secretary Royal Microscopical Society (c) ANON	(a) Lib (b) Bot lib (c) Lib	Stevens 343 lots 21pp.	J p.m.n.
February 14	(a) [HELMS (Richard)] (b) – (e) ANON	(a) European col Exotic col (b) Br lep (c) Exotic butts (d) Min HH (e) Sh	Stevens 527 lots 28pp.	J m.p.m.n.
March 28	(a) – (e) ANON	(a) Br lep (b) Ool (c) HH Sh (d) Birds (e) Lib	Stevens 356 lots 18pp.	J s.p.f.n.
April 25	(a) DICKSEE (Arthur) (b) ANON	(a) Exotic butts (b) Br lep Birds Ool	Stevens 372 lots 21pp.	J m.p.m.n.; **BMNHE**

1893 – 1894	SOURCE	CONTENTS	AUCTIONEER & SALE CAT.	REFERENCE
May 16	(a) ANON (b) FRASER (William Thomson) (c) ANON (d) DICKSEE (Arthur) (e) – (f) ANON	(a) Birds (b) Birds (c) *Aepyornis maximus* egg (d) Exotic butts (e) Exotic lep Exotic col (f) Sh	Stevens 347 lots 17pp.	J m.p.m.n.
May 26	MORRIS (Rev. F.O.), The Rectory Nunburnholme, East Yorkshire	For butts Br butts Br & For moths Col Dipt Ool Cab For arachnidae For crustacea (Starfish)	Richardson & Trotter, *house sale*	*Nature*, May 1893: xviii
June 20	(a) FARR (W.B.)† (b) [ESCOMBE, Sidcup] (c) – (d) ANON	(a) Indian ins (b) Natal lep (c) Br ool For ool (d) Birds	Stevens 344 lots 19pp.	J s.p.s.n.
July 11	[BOUCARD]	Mam Birds Ool (incl. *Aepyornis maximus*) Nh	Stevens 355 lots 16pp.	J
July 12	(a) HEULAND† (b) – (d) ANON	(a) Min (b) Br lep Birds Mam (c) Sh (d) Lib	Stevens 442 lots 22pp.	J s.p.f.n.
July 18	(a) ANON (b) [NOWERS (J.E.)] (c) [TAYLOR (Mrs.)] (d) SHEPHERD (E.), former Secretary Entomological Society	(a) Exotic lep (b) Br lep (c) Br lep (d) Lib	Stevens 402 lots 24pp.	J p.m.n.
October 10	(a) – (b) ANON	(a) Br lep Exotic lep (b) Birds Mam Sh Min HH Nh	Stevens 364 lots 14pp.	J s.p.
November 6	BURNEY (Rev. Henry)	Lib	Stevens 296 lots 24pp.	J m.p.m.n.
November 20	ANON	Ool (*Aepyornis maximus*), 1	Davis 47 lots 6pp.	**BMETH**
November 21-22	BURNEY (Rev. Henry)	Br lep (part I)	Stevens 605 lots 32pp.	J p.n.; **BMNHE; CH**
December 4	(a) ANON (b) [LARCOM (T.H.), 54 Shaftesbury Rd., Gosport] (c) – (d) ANON	(a) Br lep For lep (b) Br lep (c) Tenasserim butts (d) HH Ool Birds Mam	Stevens 404 lots 19pp.	J m.p.s.n.
1894 January 8	(a) SPILLER (A.J.) (b) – (d) ANON (e) HENSMAN (Dr.)† (f) SWINHOE (Colonel) (g) – (j) ANON	(a) Br lep Exotic lep (b) Exotic butts (c) Br lep (d) Ool (e) Br lep (f) Exotic butts (dup) (g) Br lep (h) Birds (i) HH (j) Ool	Stevens 491 lots 27pp.	J m.p.m.n.
January 22	(a) BURNEY (Rev. Henry) (b) ST. JOHN (Rev. Seymour)[1] (c) ANON	(a) Br lep (part II, incl. microlep dup macrolep) (b) Br lep (c) Natal lep	Stevens 396 lots 25pp.	**BMNHE;** J p.m.n.; **CH** s.p.; **SAR**

[1] O. Pickard-Cambridge has an interesting note (in *Entom. Record*, 5: 74) questioning the origin and authenticity of some of the specimens in both the Burney and St. John sales.

1894	SOURCE	CONTENTS	AUCTIONEER & SALE CAT.	REFERENCE
February 22	(a) BURNEY (Rev. Henry) (b) HAMONVILLE (Baron d') (c) COCKBURN (Charles)	(a) Ool (b) Dup ool (incl. Great Auk) (c) Birds Ool	Stevens 265 lots 18pp.	J m.p.m.n.; N; SAR
February 26	(a) ANON (b) SWINHOE (Colonel) (c) [ROSSER (C.W.S.), Culow House, Weston-super-Mare] (d) [STOKE-ROBERTS (Mrs.), 14 Prospect Rd., Chatham, Kent] (e) ANON	(a) Exotic lep (b) Exotic butts (dup) (c) Br lep European lep (d) Exotic butts (e) Birds Ool Exotic lep	Stevens 483 lots 23pp.	J m.p.m.n.; SAR
March 13	HENSMAN (Dr.)	Lib	Stevens 441 lots 23pp.	J
March 14	[HULKES], Little Hermitage, Higham, near Rochester, Kent.	Sh Fos Ool, 1[1]	*house sale*	Parkin (1911:15
April 24	(a) WATSON (Rev. J.), Upper Norwood† (b) ANON (c) MOORE (d) ANON (e) CLARK (S.V.) (f) HALL (George), Leicester	(a) Br lep (b) Exotic lep Indian col (c) Indian hym Orth Hem (homoptera) Dipt Br galls Exotic galls (d) Birds (*Aepyornis maximus*) Ool (incl. 2 Great Auk) (e) Dup ool (f) Birds	Stevens 549 lots 30pp.	BMNHE; J m.p. m.n.; N
May 22	(a) PAGE (W.), Smithfield and Barnes Common† (b) [PRATT] (c) ANON (d) MOORE (e) HALL (George), Belgrave, Leicester	(a) Br lep (b) Br lep (c) Pal lep (d) Indian hym Hem (homoptera) Dipt Orth Br ins Exotic ins Galls (e) Br birds	Stevens 440 lots 24pp.	J m.p.m.n.; SAR
May 28	(a) WEIR (J. Jenner)† (b) FRANCIS (Horace)†	(a) Lib (b) Lib	Stevens 255 lots 19pp.	J p.n.; BMNHE
May 29	(a) WEIR (J. Jenner)† (b) FRANCIS (Horace)† (c) WOLLASTON (J. Vernon)	(a) Br lep Exotic butts (b) Br col Br hem (heteroptera) Br hym (c) Atlantic Isles col	Stevens 313 lots 25pp.	J p.n.; BMNHE; CH; H; SAR
June 11	ANON	Lib	Stevens 385 lots 23pp.	J
June 19	(a) WEIR (J. Jenner)† (b) HEARDER (Dr. G. J.), Carmarthen	(a) Exotic lep (b) Br lep	Stevens 292 lots 23pp.	J p.n.; CH; H
June 26	(a) [BOUCARD] (b) FRANCIS (Horace)† (c) – (g) ANON	(a) Indian lep (b) Br hem (heteroptera) (c) Br col (d) Phillipines lep (e) Ool HH Birds Min Crustacea (f) Lib (g) Exotic lep	Stevens 319 + 16 bis lots 23pp. + Addenda (1 leaf)	J
July 10	WELLMAN (J.R.)	Br macrolep Br microlep	Stevens 338 lots 23pp.	CH s.p.; BMNHE H; J p.n.

[1] Included in this lot were two eggs of the Great Auk bought by Wallace Hewitt of Newington, Kent, for 36s.

1894 – 1895	SOURCE	CONTENTS	AUCTIONEER & SALE CAT.	REFERENCE
July 11	(a) 'From a celebrated French collection' (b) – (c) ANON (d) FRANCIS (Horace)† (e) – (f) ANON	(a) Birds Mam (b) Br lep Exotic lep (c) Br birds (d) Br hem (heteroptera) (e) Exotic butts (f) Birds	Stevens 275 lots 12pp.	J; SAR
August 28	(a) MAINWARING (Gen. G.B.) (b) CROUCH (Walter), F.Z.S.	(a) Sh (b) Sh	Stevens 317 + 15 bis lots 18pp.	SPD m.p.
September 25	(a) – (c) ANON	(a) Mam (b) Birds (c) For lep	Stevens 294 lots 12pp.	J; SAR
October 23	DOWNING (John W.)	Br lep	Stevens 327 lots 18pp.	J p.n.; BMNHE; CH; SAR
October 30	(a) – (c) ANON (d) BEECHER (H.M.) (e) – (h) ANON	(a) Exotic butts (b) Exotic butts (c) Sh (d) Min (e) Min (f) Fos (g) Entomological app (h) Birds	Stevens 366 lots 18pp.	J
November 20	(a) DOWNING (John W.) (b) – (c) ANON (d) RAINE (F.), Durham (e) ANON	(a) Br lep (b) For lep For col (c) Birds (d) Birds (e) HH	Stevens 371 lots 18pp.	J m.p.m.n.; CH; SAR
December 11	(a) – (d) ANON	(a) Canary Is (b) Ool (c) Fos (d) Exotic lep	Stevens 240 lots 12pp.	J; N m.p.s.n.
1895 January 14	(a) – (e) ANON	(a) Tropical lep (b) Br lep (c) Exotic lep (d) Exotic butts (e) Br col	Stevens 324 lots 17pp.	J p.m.n.
February 11	(a) [LONGLEY (W.)] (b) – (e) ANON (f) FRANCIS (Horace)† (g) ANON	(a) Entomological app (b) Lib (c) Sh (d) Min (e) Ool HH (f) Hem (heteroptera) (g) Ool	Stevens 420 lots 22pp.	J; SAR
February 21	(a) ANON (b) [MARSDEN (H.W.)] (c) ANON (d) THURNALL (Charles), Whittlesford, Cambs.† (e) DAVENPORT (H.S.), Skeffington, Leicester	(a) Br ool (b) Ool (c) Ool (*Aepyornis maximus*) (d) Ool (e) Ool	Stevens 241 lots 15pp.	J s.p.s.n.; N
February 26	MACHIN (William)†, Part I	Br lep	Stevens 298 lots 21pp.	J p.n.; CH s.p.; BMNHE; LC
March 26	(a) – (f) ANON (g) FRANCIS (Horace)†	(a) Ool (b) Ool (c) Exotic lep (d) For butts (e) Br lep (f) Min (g) Br hem (heteroptera)	Stevens 240 lots 12pp.	J s.p.s.n.
April 22-23	(a) FIELD (Leopold), Part I (b) MILNER (Sir William)	(a) Ool Nests (b) Great Auk and egg	Stevens 488 + 7 bis lots 32pp.	J p.n.; CH; N m.p.m.n.
April 30	(a) O'REILLY (The) (b) – (e) ANON	(a) Br lep (b) Exotic lep (c) Min (d) Birds (e) Ool	Stevens 307 lots 17pp.	J m.p.m.n.; CH; SAR

1895	SOURCE	CONTENTS	AUCTIONEER & SALE CAT.	REFERENCE
May 9	(a) FIELD (Leopold), Part II (b) – (c) ANON	(a) – (c) Ool	Stevens 288 lots 23pp.	J p.m.n.; N m.p.m.n.
May 21	(a) [GROOM – NAPIER, (b) – (f) ANON (g) [?PARRY] (h) [NOBLE]	(a) Fos Sh (b) Ool (c) White baboon (d) Sh ('chiefly from the China Sea') (e) Land sh (f) Exotic lep (g) Br lep (larvae) (h) Ool	Stevens 401 lots 25pp.	J m.p.m.n.
May 28	MACHIN (William)†, [Part II]	Br lep	Stevens 288 lots 20pp.	J p.m.n.; BMNHE p.n.; CH p.n. copy
June 11	(a) ROBSON (J.E.) (b) MANTUA (Duchess of)† (c) ANON	(a) Br lep (b) Exotic ins (c) Exotic lep	Stevens 333 lots 23pp.	CH f.p.; BMNHE J m.p.m.n.; LC
June 12	(a) MANTUA (Duchess of)† (b) – (c) ANON (d) BARRON (Charles), formerly Curator, Museum of Royal Naval Hospital, Haslar (e) HADLEY (E.B.), Horeham Manor, Sussex.	(a) Nh Min Crustacea Bot Sh (b) Fos (c) Fos Min (d) Birds Mam (e) Br ool	Stevens 375 lots 18pp.	J; SAR
June 25-26	(a) FIELD (Leopold)†, Part III (b) HAMONVILLE (Baron d') (c) ANON	(a) Ool (b) Ool (Great Auk) (c) Ool	Stevens 567 lots 37pp.	J p.n.; N m.p. m.n.
July 9	(a) MANTUA (Duchess of)†[1] (b) – (d) ANON	(a) Sh Herb (incl. Herb DESEGLISÉ) (b) Lib (c) Sh Min Exotic butts (d) Sh Fos	Stevens 499 lots 22pp.	J
July 10	WHEELER (F.D.), Norwich	Br lep	Stevens 300 lots 18pp.	J p.n.; CH s.p.; BMNHE
July 16	(a) ANON (b) HARKER (Professor Allen) (c) – (d) ANON (e) MANTUA (Duchess of)† (f) ANON (g) MANTUA (Duchess of)†[2]	(a) Br lep (b) Br ins (incl. col hem dipt hym) Exotic lep Exotic col (c) Exotic butts (d) Br butts (e) Br lep Exotic lep Col (f) Br lep (g) Sh Fos Mam Herb	Stevens 392 lots 19pp.	J m.p.m.n.; SAR inc.
July 31	(a) MANTUA (Duchess of)† (b) PRUEN (J.A.) (c) ANON	(a) Sh Fos Herb (b) Ool (c) Ool	Stevens 200 lots 13pp.	J
October 14	ANON	Fos Min Birds Nh Irr	Stevens 393 + 6 bis lots 16pp.	BMETH f.p.
October 22	(a) ROBSON (J.E.) (b) – (c) ANON	(a) Br macrolep Br microlep (b) Exotic lep (c) Lib	Stevens 272 + 3 bis lots 17pp.	J p.n.; CH; BMNHE
November 11	(a) WHEELER (F.D.), Norwich (b) – (c) ANON	(a) Br macrolep Br microlep (b) For lep (c) Birds Ool	Stevens 295 + 4 bis lots 22pp.	J m.p.m.n.; BMNHE; CH; LC

[1] Included were a number of large herbaria and seventeen vols. of MS catalogues.

[2] The wife of Chas. Ottlery Groom, 'a notorious rogue and thief, [who] tried to kill Thomas Davies by dropping a boulder upon him from a high ladder in Tennant's shop in the Strand.' Groom 'became successively, Groom - Napier, [see his sale of 21.v.1895] Duke of Mantua and Monteferrat, Prince of Mantua, Prince of the House of David' (cf. Sherborn, 1940:61).

1895 – 1896	SOURCE	CONTENTS	AUCTIONEER & SALE CAT.	REFERENCE
November 19	(a) FARREN (Wm.), Cambridge, Part I (b) CURTIS (John)†	(a) Br macrolep Br microlep (b) Lep	Stevens 332 + 1 bis lots 21pp.	J p.n.; BMNHE; CH; LC
December 2	(a) FARREN (Wm.) Cambridge, Part II (final) (b) ATKINSON (E.T.) (c) – (e) ANON	(a) Microlep (b) Indian ins (c) Exotic lep (d) Birds Ool Min Fos Sh Nh (e) Lib	Stevens 420 + 18 bis lots 24pp.	J p.n.; CH; BMNHE; LC
December 9	ANON	Br ool	Stevens 280 lots 31pp.	J m.p.m.n.; SAR
December 10	TUGWELL (W.H.)†, Part I	Br lep	Stevens 295 lots 21pp.	J p.n.; CH p; LC; BMNHE p.m.n.; SAR
December 16	ANON	Mam HH Sh Birds Nh, 41	Stevens 298 + 25 bis lots 14pp.	BMETH
1896 January 20	(a) TUGWELL (W.H.), Part II (final) (b) BARTLETT (F.) (c) – (d) ANON (e) ROSS (H.)† (f) LEWCOCK (G.A.)	(a) Br macrolep Br microlep (b) Br lep (c) Herb (d) Br col (e) Columbian ins (f) Col	Stevens 256 + 2 bis lots 21pp.	CH m.p.; J m.p.m.n.; BMNHE m.p. m.n.; SAR
January 27	(a) LEIPNER (Professor) (b) – (e) ANON	(a) Sh (b) Fos Sh (c) Exotic butts (d) Lep and Col collected in Florida in 1895 (e) Br lep	Stevens 275 + 52 bis lots 15pp.	BMETH; J m.p. m.n.
February 17	(a) RUSS (P.), Sligo (b) ANON	(a) Br lep (b) Exotic lep Ool Nh	Stevens 390 lots 20pp.	CH m.p.; BMNHE; J m.p. m.n.
March 9-10	(a) FRY (C.E.) (b) STANDISH	(a) Br macrolep Br microlep (b) Microlep	Stevens 593 lots 24pp.	J p.n.; BMNHE; CH
March 16	(a) – (d) ANON	(a) Exotic butts (b) Exotic butts (c) Ool Crustacea Br fos (d) Sh	Stevens 319 lots 16pp.	J f.p.f.n.; SAR
April 13	(a) SLADEN (C.A.), incl. his father E.H.M.SLADEN coll. (b) – (c) ANON	(a) Br lep (b) Exotic lep (c) Ool Min Fos Nh	Stevens 409 lots 23pp.	J m.p.m.n.; BMNHE
April 20	(a) TUKE (J.H.) (b) – (c) ANON (d) MANTUA (Duchess of)† (e) – (f) ANON	(a) Ool (Great Auk) (b) Ool (*Aepyornis maximus*) (c) Ool (*Aepyornis grandidieri*) (d) Ool (part I) (e) Br ool (f) Ool	Stevens 279 lots 19pp.	J f.p.f.n.; N
May 4	TYLER (Capt. Charles)†	Lib	Stevens 297 lots 19pp.	J s.p.s.n.
May 8	ANON	Microscopes Microscopic nh specimens Irr	Stevens 516 lots 20pp.	J
May 18	(a) YOUNG (S.), Cambridge (b) [FRERE (Rev. E.H.)] (c) – (f) ANON	(a) Br lep (b) Br lep (c) Exotic ins (d) Nh (e) Sh (f) Nh Min	Stevens 388 + 12 bis lots 20pp.	J m.p.m.n.; CH

1896 – 1897	SOURCE	CONTENTS	AUCTIONEER & SALE CAT.	REFERENCE
May 22	TYLER (Capt. Charles)†	Microscopes Microscopic nh specimens Irr	Stevens	Advert on p.20 of Stevens' Sale Cat. of 18 v 1896
June 15	(a) WILLIAMS† (b) TYLER (Capt. Charles)† (c) TUCKER (Marwood), Coryton Park, Axminster, Devon (d) ANON	(a) Br lep (b) Fos Min Nh (c) Birds Min Fos Nh (d) Ins Sh Nh	Stevens 442 + 24 bis lots 24pp.	BMNHE; J s.p. s.n.; CH f.p.
June 16	(a) MANTUA (Duchess of)† (b) DAVENPORT (H.S.) (c) – (d) ANON	(a) – (c) Ool (d) Ool (*Aepyornis grandidieri*)	Stevens 290 lots 17pp.	J m.p.s.n.
July 8	(a) COOPER (J.A.) (b) – (c) ANON	(a) Br macrolep Br microlep (b) Exotic lep Exotic col (c) Fos Min Nh	Stevens 349 + 5 bis lots 20pp.	J m.p.m.n. BMNHE; CH s.p SAR inc.
July 21	(a) COOPER (J.A.) (b) – (d) ANON	(a) Ool (b) Ool (c) N.American ool (d) Ool	Stevens 358 lots 26pp.	J
August 18	(a) LEEDS (Alfred N.), Eyebury, Peterborough (b) ANON (c) MANTUA (Duchess of)†	(a) Fos (from Oxford Clay of Fletton, Hunts.) (b) Exotic lep Exotic col (c) Min	Stevens 205 lots 11pp.	J
September 22	(a) [PRUEN?] (b) ANON (c) MANTUA (Duchess of)† (d) ANON	(a) Exotic lep (b) Exotic lep Br lep (c) Min (d) Sh	Stevens 329 lots 17pp.	J f.p.f.n.
October 27-28	BRIGGS (C.A.), Part I	Br macrolep Br microlep	Stevens 457 + 12 bis lots 29pp.	J p.n.; CH m.p.; BMNHE; LC
November 2	(a) – (d) ANON	(a) Exotic butts Entomological app (b) Ool (c) Sh (d) HH Mam	Stevens 306 lots 15pp.	J f.p.f.n.; SAR
November 10	(a) BRIGGS (C.A.), Part II (b) SMART (J.) (c) GRIFFITH (Rev.), Stratfield Turgiss (d) SMITH (James), Plumstead	(a) Br macrolep Br microlep (b) Br ins For ins (c) Br lep Br col (d) Br lep For lep	Stevens 281 lots 20pp.	J p.m.n.; LC BMNHE; CH
November 24-25	BRIGGS (C.A.), Part III (final)	Br macrolep Br microlep	Stevens 481 + 9 bis lots 29pp.	J p.n.; BMNHE; CH s.p.
December 8	(a) – (e) ANON	(a) Exotic butts (b) Lep (c) Exotic lep (d) HH (e) Birds Sh Ool	Stevens 266 lots 12pp.	J; SAR
1897 January 25	(a) – (c) ANON	(a) Exotic lep (b) HH (c) Min	Stevens 251 lots 35pp.	J
February 22	(a) – (d) ANON	(a) Min (b) Exotic lep (c) Br lep (d) Mam	Stevens 404 lots 18pp.	J
March 15	CHOLMONDELEY (Reginald), Condover Hall, Shrewsbury†	Sh	Stevens 236 + 1 bis lots 10pp.	NMW p. (incl. J.C.MELVILL's purchases); J

1897	SOURCE	CONTENTS	AUCTIONEER & SALE CAT.	REFERENCE
March 22	(a) – (e) ANON	(a) Exotic butts (b) Lep (c) Br lep For lep (d) Br lep (e) Shetland lep	Stevens 348 lots 20pp.	J m.p.m.n.
April 13	(a) SMART (Rev. Gregory)† (b) – (c) ANON (d) HOPKE (e) FIELD (Leopold)†	(a) Ool (b) Ool (Great Auk) (c) Ool (d) Chilean ool (e) Ool	Stevens 389 lots 28pp.	J s.p.s.n.
April 27	CHOLMONDELEY (Reginald), Condover Hall, Shrewsbury†	World ins (lep col orth hem neur)	Stevens 301 lots 20pp.	J m.p.m.n.; SAR
May 4	(a) CHOLMONDELEY (Reginald), Condover Hall, Shrewsbury† (b) – (c) ANON (d) TOUCH (J. de la) (e) GRAHAM-CLARKE (L.T.) (f) – (h) ANON	(a) Birds (b) Sh Min (c) HH (d) Birds from China (e) Birds from Ceylon (f) HH (g) Exotic lep (h) Exotic butts	Stevens 299 lots 15pp.	J s.p.s.n.
May 17	ANON	Lib	Stevens 219 lots 15pp.	J
May 18	STUART (Mrs. W.), Aldenham Abbey, Aldenham, Watford†	Min Fos Cor	Stevens 189 lots 11pp.	**BMNHM**
May 25	ANON	HH Irr	Stevens	Advert on p.11 of Stevens' Sale Cat. of 18 v: 1897
June 2	(a) BROAD (C.) (b) – (c) ANON	(a) Br lep Exotic lep Exotic col (b) Exotic lep (c) HH	Stevens 375 lots 20pp.	J m.p.m.n.
July 27	(a) CHAMPLEY (Robert), Scarborough† (b) – (d) ANON	(a) Ool (b) Ool (Great Auk) (c) Ool (d) Birds	Stevens 280 lots 17pp.	J
August 24	(a) ANON (b) [CREGOE (J.P.), Tredinick, Mayon Rd., Sydenham London, S.E.] (c) – (e) ANON	(a) Birds (b) Dup S.African ins (lep col hym orth) (c) Min Fos (d) For lep (e) HH	Stevens 295 lots 15pp.	J p.n. of (b)
September 28	(a) – (c) ANON	(a) Lib (b) Exotic lep (c) HH	Stevens 393 lots 20pp.	J f.p.f.n.; SAR
November 8	CALVERT (John)†, Part I	Nh	Stevens	*Athenaeum*, Oct., **1897**:543
November 22-23	HODGKINSON (J.B.)†, Part I	Br macrolep Br microlep	Stevens 538 lots 30pp.	J p.n.; **BMNHE** p.m.n.; **CH** s.p.; SAR
November 29	CALVERT (John)†, Part II	Nh Irr	Stevens	Advert on p.30 of Stevens' Cat. of an irr. sale of 22-23 ix 1897 in BMETH
December 6	(a) ASHBY (Richard), Egham† (b) – (c) ANON	(a) Birds Sh Min (b) New Zealand birds incl. skeleton of Moa (c) HH	Stevens 436 lots 21pp.	J f.p.f.n.; SAR

1897–1898	SOURCE	CONTENTS	AUCTIONEER & SALE CAT.	REFERENCE
December 14	(a) HODGKINSON (J.B.)†, Part II (final) (b) MATTHEWS (Rev. A.), Gumley (c) HERVEY (Rev. A.C.) (d) – (f) ANON	(a) Br microlep (b) Br lep Br col (c) Br lep (d) Lib (e) Exotic lep (f) Ins Nh	Stevens 450 + 3 bis lots 26pp.	CH m.p.; LC; J p.m.n.; BMNHE s.p.s.n.; SAR
December 20	(a) ANON (b) HEYDE (Rev. Thos. H.) (c) – (d) ANON	(a) Birds Ool (b) Birds from Panama & Guatemala (c) Sh (d) Mam	Stevens 351 lots 17pp.	J s.p.s.n.
1898 February 7	(a) – (b) ANON (c) BAYNES (Sir William John Walter)† (d) ANON	(a) Exotic lep (b) Birds Ool (c) HH (d) HH	Stevens 293 lots 12pp.	J s.p.s.n.
February 12	(a) OSBORNE (A.P.) (b) [COOKE]	(a) Nh (b) Birds Ool Lep Col HH Lib Nh Irr	Bullock 291 lots 14pp.	J p.s.n. incl. all (b)
March 14	(a) HAWKINS (Rev. Hubert H.), Beyton† (b) PATTERSON (George), Bearsden, Glasgow† (c) NOBLE (H.) (d) ANON	(a) Br ool (b) Br ool (c) Dup for. ool (d) For ool	Stevens 325 lots 23pp.	J s.p.s.n.; N f.n. SAR
March 17	HEYWOOD (James), F.R.S., F.S.A., F.G.S.†	Fos, 2	Stevens 546 lots 28pp.	BMETH
March 21	(a) MATTHIAS (H.W.), Thames Ditton (b) – (d) ANON	(a) Br lep (b) Exotic lep (c) HH (d) Sh	Stevens 314 lots 18pp.	J m.p.m.n.; CH; BMNHE
April 4	ANON	Min Fos, 4	Stevens 398 lots 25pp.	BMETH
April 18	(a) [DONCASTER] (b) – (e) ANON	(a) For lep (his stock of) (b) Br macrolep (c) HH (d) Ool (e) Exotic lep	Stevens 372 lots 20pp.	J m.p.
April 25	ELISHA (George)	Br macrolep Br microlep	Stevens 291 lots 22pp.	J p.n.; CH f.p.; BMNHE; SAR inc.
May 24	(a) HOLLIS (G.) (b) SOUTH (R.) (c) ELISHA (George) (d) ANON (e) BARCLAY (Sir David) (f) ANON	(a) Br lep (b) Br microlep (c) Dup Br lep (d) Exotic lep (e) Sh (f) Min	Stevens 368 lots 21pp.	J m.p.m.n.; BMNHE; CH; SAR
July 4-7	CONNOP (Mrs. Arthur), Bradfield Hall, near Reading	4 vii: Birds Nh, 60	Robinson & Fisher, *house sale* 797 + 4 bis lots 40pp.	CLD
July 5-6	ANON	Mam Nh, c.30	Stevens 510 + 27 bis lots 30pp. + 2 plates	BMETH
July 19-20	(a) BANKS (Miss), St. Catherine's, Doncaster† (b) CALVERT (John)†, Part V (c) ANON (d) PRUEN (J.A.) (e) ANON	(a) Cor Sh Fos Min Cab Lib (b) Birds Ool Nests (c) HH Exotic butts (d) Br lep (e) Birds	Stevens 793 lots 32pp.	J s.p.s.n.; SAR

1898 – 1899	SOURCE	CONTENTS	AUCTIONEER & SALE CAT.	REFERENCE
October 17	ANON	Exotic lep	Stevens 398 lots 16pp.	J f.p.f.n.
December 5	(a) SOUTH (R.) (b) – (f) ANON	(a) – (b) Br macrolep Br microlep (c) Lep (d) Ins Nh (e) Min (f) Sh	Stevens 428 + 30 bis lots 27pp.	J; CH; BMNHE; SAR
1899 February 13	(a) [SHIRLEY] (b) – (c) ANON (d) SHIRLEY, Effington (e) – (f) ANON	(a) Br lep (b) Exotic lep (c) HH Ins Nh (d) Birds (dup) (e) Mam Ins Nh (f) Ool Nh	Stevens 415 + 16 bis lots 21pp.	CH; J; SAR
March 13	(a) ANON (b) [WEBB (S.)] (c) – (d) ANON	(a) Exotic lep (b) Lep (mimics) (c) Exotic lep (d) Sh	Stevens 342 lots 19pp.	J s.p.s.n.; SAR
March 27-28	(a) BARTON (Stephen), Bristol† (b) ANON	(a) World col (b) Exotic lep Br lep Br col Lib	Stevens 676 lots 44pp.	J p.n.; SAR
April 17	(a) [DONCASTER] (b) ANON (c) [SWINHOE] (d) – (e) ANON	(a) Exotic col (b) Exotic butts (c) Exotic lep (d) HH (e) Sh	Stevens 462 lots 20pp.	J s.p.s.n.; SAR
April 19	STAINTON (H.T.), incl. J.F.STEPHENS (Part of)	Lib	Sotheby 225 lots 19pp.	BMNHE; J
May 16	(a) HADFIELD Newark† (b) [CROTCH (G.R.)] (c) [SCHILL & LOCKE] (d) – (f) ANON	(a) Br lep Br col (b) Br col (dup) (c) Lep Col (d) Br macrolep Br microlep (e) HH (f) Sh Min Fos	Stevens 455 lots 20pp.	J m.p.m.n.
May 30	(a) WHITELY, Woolwich (b) HADFIELD, Newark† (c) ETON COLLEGE MUSEUM (mostly from coll. of Provost THACKERAY of Kings College, Cambridge) (d) NOBLE (Heatley) (e) – (f) ANON	(a) Humming birds (b) Birds (c) Dup birds (d) Birds (e) Br birds (f) Sh Fos Min	Stevens 352 lots 15pp.	J
June 13	LANGDON (A.W.)†	Sh	Stevens 235 lots 15pp.	NMW s.p. (J.C. MELVILL's purchases)
June 21	(a) – (c) ANON	(a) HH (b) Min (c) Br macrolep	Stevens 383 lots 20pp.	J f.p.f.n.; SAR
July 18	(a) HAWKE (Wm.)† (b) – (f) ANON	(a) Exotic lep (b) Br lep (c) Br lep (d) Lib (e) Australian fish (f) HH	Stevens 358 lots 19pp.	J s.p.s.n.
July 19	(a) FIELD (Leopold)† (b) NOBLE (Heatley) (c) HAMMONVILLE (Baron d') (d) ANON	(a) Ool (b) Dup ool (c) Ool (Great Auk) (d) Birds Ool	Stevens 250 lots 23pp.	J m.p.m.n.
August 22	(a) ANON (b) LEES, Clarksfield (c) – (d) ANON	(a) Exotic birds (b) Ool (c) HH (d) Min Fos	Stevens 270 lots 12pp.	J s.p.s.n.; N

135

1899 – 1900	SOURCE	CONTENTS	AUCTIONEER & SALE CAT.	REFERENCE
September 19	(a) – (e) ANON	(a) Min Fos (b) Birds (c) HH (d) Ool (e) Mosses	Stevens 361 lots 18pp.	**J** s.p.s.n.; **N**; SAR
October 24	(a) – (e) ANON	(a) Br lep (b) Exotic lep (c) Br lep (d) Birds Ool (e) HH	Stevens 261 lots 14pp.	**J** f.p.f.n.; SAR
November 20	(a) MERRIN (Joseph) (b) – (f) ANON	(a) Br lep (b) Exotic lep (c) HH (d) Mam (e) Tortoises & Turtles (f) For ool	Stevens 359 lots 19pp.	**BMNHE**; **J** s.p.s.n.; **CH**
December 18	(a) LOVELL-KEAYS (A.)† (b) – (f) ANON	(a) Br lep (b) Exotic lep Exotic col (c) Birds Mam (d) HH (e) HH (f) New Guinea birds	Stevens 431 lots 23pp.	**J** s.p.s.n.; **BMNHE**
1900 January 23	(a) LABREY (B. Bowman)† (b) – (e) ANON (f) WOLLASTON (T. Vernon)	(a) Exotic butts (b) Exotic lep (c) HH (d) Birds Ool (e) Sh (f) 'The Collection of Land and Freshwater Shells from Madeira'	Stevens 298 lots 14pp.	**J** f.p.f.n.; **H**; **BMNHL** copy
February 19	(a) – (e) ANON (f) ASHMOLEAN MUSEUM, OXFORD (g) RAWSON (Sir H. Rawson) (h) – (l) ANON	(a) Mam (b) Birds Ool Nests (c) Tortoises (d) Scottish ool (e) Sh (f) Lib (g) Sh (h) Br lep (i) Br lep For lep (j) – (k) Exotic lep (l) Birds HH	Stevens 471 lots 26pp.	**J** m.p.m.n.
March 19	(a) – (f) ANON	(a) Br lep For lep (b) Exotic lep (c) HH (d) Sh (e) Borneo sh (f) Birds Ool Nests	Stevens 412 lots 18pp.	**J** f.p.f.n.; SAR
March 27-28	STEVENS (Samuel)† Part I	Br macrolep Br microlep	Stevens 553 lots 29pp.	**J** p.n.; **CH** m.p. **GS**; **BMNHE**; SAR
April 23	STEVENS (Samuel)† Part II (final)	Br macrolep Br microlep	Stevens 323 lots 22pp.	**J** p.n.; **CH**; **BMNHE**; SAR
May 1	[SWINHOE, MARSDEN & JEFFREYS]	Exotic lep Br lep HH Birds Lib	Stevens 419 lots 23pp.	**J** m.p.m.n.; SAR
May 29	BRIGHT (P.M.), Bournemouth	Br macrolep	Stevens 335 lots 17pp.	**J** p.n.; **CH** p.; **GS**; **BMNHE**; H SAR; MAR p.n.
June 19	(a) BRIGHT (P.M.), Bournemouth (b) CROTCH (W. Duppa) (c) LOWE (W.H.), Wimbledon	(a) Br macrolep Br microlep (b) Br lep Col (c) Br lep	Stevens 319 lots 20pp.	**J** p.m.n.; **GS**; **BMNHE**; H; SAR
June 20	(a) MUNT (H.) (b) ANON (c) FURNEAUX (Rev. A.) (d) ANON (e) ROSOMON (H. jun.)	(a) Br ool (b) Ool (two Great Auk) (c) Ool (d) Br ool Nests (e) Ool	Stevens 411 lots 26pp.	**J** m.p.m.n.
June 27	(a) VIRTUE-TEBBS (H.)† (b) – (d) ANON	(a) Min Fos Birds (skeletons) Mam (skeletons) Nh (b) Exotic lep (c) Exotic col (d) Br lep	Stevens 340 lots 20pp.	**J** m.p.m.n.
July 24	(a) – (e) ANON	(a) Sh (b) Min Fos (c) Br lep (d) HH (e) Ool	Stevens 531 lots 27pp.	**J**; **NMW**; SAR

1900 – 1901	SOURCE	CONTENTS	AUCTIONEER & SALE CAT.	REFERENCE
October 22	(a) – (e) ANON	(a) Exotic lep (part I) (b) Exotic lep (c) Br lep (d) Exotic col (e) Exotic lep	Stevens 346 lots 19pp.	J m.p.m.n.; CH
November 12	(a) ANON (b) [KEAYS (Lovell)]	(a) Br lep Exotic lep (b) Br lep	Stevens 349 lots 22pp.	J p.n.; BMNHE; CH inc.; SAR
November 20	(a) WALLIS (S.H.) (b) – (c) ANON (d) LEWIS (H.), Clapham† (e) – (f) ANON	(a) Ool (b) HH Mam bones (c) HH Nh (d) Fos Nh (e) Nh (f) Ins Exotic lep Br lep North American butts.	Stevens 349 lots 16pp.	J s.p.s.n.; BMNHE; SAR
December 18	(a) NEWAN (W.), Darlington† (b) FIELD-FISHER (Thomas), Cambridge (c) – (d) ANON	(a) Br lep (b) Br lep (c) Exotic lep (d) Ool HH Nh	Stevens 480 lots 23pp.	BMNHE; SAR
1901				
January 29	(a) GARDINER (Mrs.), St. John's Wood Park (b) – (h) ANON	(a) Sh Cor (b) Min Fos (c) HH (d) Lib (e) Sh f Crustacea (g) Mam (h) Exotic col Exotic lep	Stevens 483 lots 22pp.	J f.p.f.n.; SPD; SAR
March 4	(a) – (g) ANON	(a) S.African hh (b) Birds (c) Exotic lep (d) Entomological app (e) Br lep (f) Sh (g) Sh	Stevens 429 lots 22pp.	J s.p.f.n.; SAR
March 26	(a) ABBOTT (P.W.), Edgbaston (b) ANON	(a) Br lep (b) Pal lep Col (Histeridae)	Stevens 263 lots 16pp.	CH p.; BMNHE; GS; SAR
April 15	(a) CROWLEY (Philip), Waddon Ho., Croydon (b) ANON	(a) – (b) Lib	Stevens 234 + 25 bis lots 14 + 4 pp.	J p.n.; BMNHE; N f.p.
April 30	(a) – (f) ANON	(a) Br lep (b) For lep (c) HH (d) Birds Mam (e) Fos Min Sh (f) 80 human skulls from New Guinea	Stevens 377 lots 18pp.	J p.m.n. (lep only) SAR
May 15	(a) HORLEY (W.J.), Hoddesdon (b) ANON (c) NOBLE (Heatley) (d) BLADEN (W. Wells) (e) MASSEY (H.) (f) – (h) ANON	(a) Ool (b) Ool (*Aepyornis maximus*) (c) Dup ool (d) Ool (e) Dup ool (e) – (g) Ool	Stevens 324 lots 20pp.	J s.p.s.n.
June 4	(a) LEACH (H.)† (b) MÖLLER (O.)† (c) – (h) ANON	(a) Lep from China & Japan (b) Sikkim moths (c) Exotic lep (d) HH (e) Ferns Seaweeds[1] (f) Mam HH (g) Birds (h) Sh	Stevens 318 lots 20pp.	J m.p.m.n.
July 16	(a) CROMPTON (Sydney) (b) DORMER (Lord)† (c) MATTHEWS (Rev. A.)† (d) – (i) ANON	(a) Br lep (b) Exotic col (Cicindelidae) (c) Br birds (d) For lep (e) HH (f) Birds Ool (g) HH (h) HH (i) Sh	Stevens 528 lots 27pp.	J m.p.m.n.; BMNHE; SAR

[1] Large portfolio of New Zealand ferns; three parcels of New Zealand seaweeds – from Lady BAKER'S sale at Drinkstone, Suffolk.

1901–1902	SOURCE	CONTENTS	AUCTIONEER & SALE CAT.	REFERENCE
October 1	(a) ANON (b) ESTRIDGE (H.W.) (c) – (k) ANON	(a) Sh Fos Min (b) Fish Sh Cor Crustacea (c) Fos Sh (d) HH (e) Birds (f) Indian butts (g) Exotic butts (h) Herb (i) New Zealand lep Haiphong lep (j) Exotic col (k) Lep	Stevens 374 lots 20pp.	J m.p.m.n.; SAR
October 28	YOUNG (John)†	Ornithological lib	Stevens 204 lots 18pp.	J p.n.; SAR
October 29	(a) [WALLIS (S.H.)] (b) ANON (c) SHIRLEY (S.E.) (d) YOUNG (John) (e) ATKINSON (Wm.)	(a) Ool (b) Ool (Great Auk) (c) Ool (d) Ool (e) Birds	Stevens 345 lots 20pp.	J s.p.s.n.; N m.p f.n.
December 3	(a) – (k) ANON	(a) Lep (b) Br lep (c) Exotic lep (d) Indian butts (e) Exotic butts (f) Lep (g) HH (h) Ool (i) Ool (j) Sh (k) HH Birds Min Crustacea Sh	Stevens 479 lots 27pp.	J f.p.f.n.; SAR
1902 N.d.	LIVETT (H.W.)	Br lep Br col Exotic lep Exotic col	Stevens	Horn & Kahle (1935:157)
January 21	(a) WILKINSON (S.J.) (b) MARSHALL (Rev. F.) (c) BRACKENBURY (Rev. E.B.) (d) GEDDES (Capt. G.), Toronto (e) – (f) ANON	(a) – (c) Br lep (d) North American ins (e) Exotic lep (f) Sh	Stevens 426 lots 30pp.	J m.p.m.n.; **CH**; **BMNHE**; SAR
January 28	RYDER (G.R.)†	Lib	Stevens 296 lots 21pp.	J s.p.s.n.
March 18	(a) CROWLEY (Philip), Waddon House, Croydon†[1] (b) ANON	(a) Lep (b) Col Orth Ins	Stevens 352 lots 26pp.	J p.n.; **CH** f.p.n. **BMNHE**; SAR
March 25	(a) – (h) ANON (i) ABBOT (Rev. Wilfred H.), late of Collingwood Bay (j) ANON	(a) Ins (incl. hym col dipt neur) Exotic lep (b) Birds (c) Sponges Cor (d) HH (e) Mam Birds (f) Min Fos (g) HH (h) Sh (i) New Guinea sh (j) Sh	Stevens 436 lots 19pp.	J s.p.s.n.; SAR
April 14	(a) ORMEROD (Miss E.A.), St. Albans (b) – (c) ANON	(a) Lib (b) Br lichens For lichens (c) Lib (on lichens mosses bot)	Stevens 288 lots 20pp.	J m.p.m.n.
April 15	CROWLEY (Philip), Waddon House, Croydon†	Exotic lep European lep	Stevens 380 lots 26pp.	J m.p.m.n.; **CH** m.p.; **BMNHE**; SAR
April 17	CROWLEY (Philip), Waddon House, Croydon†	Ool (*Aepyornis maximus* and Great Auk) Great Auk (skin & bones)	Stevens 264 lots 21pp.	J m.p.m.n.; N; SAR

[1] According to Allingham (1924:140) Crowley's will enabled the Brit.Mus. (N.H.) to select the best from his collections prior to the sales.

1902	SOURCE	CONTENTS	AUCTIONEER & SALE CAT.	REFERENCE
May 6	(a) Da Costa (S.J.) (b) – (c) ANON (d) HISLOP (Robert) (e) – (f) ANON (g) CHANDLER (Rev. J.B.), Witley, Surrey (h) ANON (i) [TRAVERS] (j) – (m) ANON	(a) Sh (b) Exotic lep (c) Exotic lep (d) New Guinea lep (e) Exotic lep (f) HH (g) Br birds Br ool (h) Birds Ool (i) Birds (j) Birds (k) Birds (l) Sh (m) S.Australian seaweeds	Stevens 519 lots 23pp.	J f.p.f.n.; SAR
May 8	(a) CROWLEY (Philip), Waddon House, Croydon† (b) LEECH (J.H.)	(a) For lep (b) Chinese butts Japanese butts	Stevens 302 lots 23pp.	J m.p.m.n.; CH; BMNHE; SAR
May 15	(a) CROWLEY (Philip), Waddon House, Croydon† (b) ANON	(a) Ool (b) Ool (Great Auk)	Stevens 251 lots 23pp.	J m.p.m.n.; N p.m.n.
June 5	(a) CROWLEY (Philip), Waddon House, Croydon† (b) NOBLE (H.) (c) – (e) ANON (f) MUNT (H.)	(a) – (b) Ool (c) Ool (*Aepyornis maximus*) (d) Ool (Moa) (e) Ool (Great Auk)[1] (f) Ool	Stevens 262 lots 24pp.	J m.p.m.n.; N
June 19	STARK (Dr. A.L.)	Birds Ool Ornithological lib	Stevens 217 lots 22pp.	J s.p.s.n.; N
June 24	(a) SHIRLEY (S.E.) (b) – (i) ANON	(a) Birds (b) Sh (c) Birds Mam (d) HH (e) Birds Ool (f) Exotic lep (g) Br lep (h) Br lep Exotic lep Br col (i) Br macrolep Br microlep Br col	Stevens 447 lots 22pp.	J f.p.f.n.; CH; SAR
July 29	(a) – (k) ANON (l) DOUGLAS-FOX (J.) (m) – (n) ANON	(a) Lib (b) Birds (c) Canadian birds (d) Canadian mam (e) Sh Fos (f) Birds (g) Exotic lep (h) HH (i) Br lep (j) Br macrolep Br microlep (k) Fos from Table Cape and Tasmania incl. co-types described by G.B.PITCHARD (l) Br col (m) Br lep (n) Br lep Br col.	Stevens 537 lots 27pp.	J
September 30	(a) – (d) ANON (e) HEYNE (Alexander) (f) – (h) ANON (i) WILKINSON (Rev. C.) (j) ANON	(a) Min (b) Lib (c) HH (d) Birds (e) Herb (f) Exotic lep (g) Exotic lep (h) Pal butts (i) Br lep (j) Exotic lep	Stevens 409 lots 25pp.	J m.p.m.n.
October 21	(a) USSHER (Richard J.), Cappagh House, Co. Waterford (b) – (e) ANON	(a) – (e) Ool	Stevens 400 + 26 bis lots 26pp.	N
November 18	(a) – (h) ANON	(a) Exotic lep (b) Exotic lep (c) Lep (from Aberdeen) (d) Exotic lep (e) Australian butts (f) Birds (g) African hh (h) Lib	Stevens 501 lots 23pp.	J m.p.m.n.; SAR
December 2	(a) WALLIS (Ross) (b) – (f) ANON	(a) – (b) Ool (c) Birds (d) Ool (e) HH (f) Exotic lep	Stevens 402 lots 23pp.	J s.p.s.n.; SAR

[1] A MS note in the Janson cat. states that the Great Auk's eggs were not included in the sale.

1903	SOURCE	CONTENTS	AUCTIONEER & SALE CAT.	REFERENCE
1903 January 20	(a) – (f) ANON	(a) Mam skulls (b) African hh (c) – (e) Ool (f) Lib	Stevens 355 lots 23pp.	J s.p.s.n.; SAR
January 27	(a) WILLIAMSON (J.B.)† (b) – (c) ANON	(a) Br lep (b) Exotic lep illustrating mimicry (c) Ool HH Mam skulls	Stevens 432 lots 24pp.	J m.p.m.n. (of lep) **BMNHE**; SAR
February 17	(a) WILKINSON (Rev. C.), Toft Rectory, Lincolnshire (b) – (e) ANON (f) RÉMY (Dr.) (g) – (h) ANON	(a) – (d) Ool (c) Brazilian ins (lep & col) (f) Birds Mam (g) Exotic lep (h) Br lep	Stevens 367 lots 17pp.	J f.p.f.n.; SAR
March 17	(a) – (d) ANON (e) MAY (J.W.)† (f) – (o) ANON	(a) – (b) Ool (c) Indian ool (d) Mam Birds (e) Br col (f) Br hym (g) Hym (incl. Ichneumonidae & Tenthridinidae) (h) Br col (i) Br microlep (j) Br ins (k) Br col Br hym (l) Br hym Exotic hym (m) Exotic lep (n) Lep (mimics) (o) Sh	Stevens 540 lots 31pp.	J s.p.s.n.; SAR
April 28	(a) – (h) ANON	(a) Indian ool (b) Nh (c) Indian rept Nh (d) Lib (e) Br lep (f) Exotic lep (g) HH Mam Nh (h) Rept Fish Nh	Stevens 396 lots 20pp.	**BMNHE** p.s.n.; J f.p.f.n.; SAR
May 5	CLAREMONT (Dr. C.C.), Camden Town†	Sh Fos Cor	Stevens 304 lots 14pp.	NMW; SAR
May 19	(a) FRECKELTON (Rev. T.W.), Northampton† (b) – (g) ANON	(a) Fos Min (b) Fos (c) HH (d) Mam Birds (e) Indian ool (f) Ool (g) Min Fos Sh	Stevens 433 lots 20pp.	J; SAR
June 9	(a) – (b) ANON (c) HARRIS (Dr. R. Hamlyn)	(a) Br lep (b) Exotic lep (c) Br hym For hym	Stevens 377 lots 22pp.	J m.p.m.n.; **BMNHE**; SAR
June 23	BIDWELL (Edward), Part I	Ool	Stevens 280 lots 24pp.	J m.p.m.n.; N m.p.
June 30	(a) RAWSON (Sir Rawson)† (b) – (h) ANON	(a) Sh (b) Br birds (c) Ool Nests (d) Indian ool (e) Ool (f) HH Mam (g) Min (h) Birds	Stevens 434 lots 22pp.	NMW; J
July 7	ANON	Lib	Stevens 206 lots 16pp.	J s.p.s.n.
July 21	(a) HILL (Ainsley)† (b) – (d) ANON	(a) Br lep (b) Exotic lep (c) HH (d) Indian ool	Stevens 451 lots 20pp.	J m.p.m.n.; **BMNHE**
September 1	(a) SHEPPARD (Edward)† (b) ANON (c) HARRIS (Dr. R. Hamlyn) (d) – (g) ANON	(a) For col Br col (b) Exotic lep (c) Br hym Continental hym (d) Lep (e) HH Mam (f) Indian ool (g) Sh	Stevens 403 lots 23pp.	J p.n. of (a); N
September 29	(a) – (c) ANON (d) WHITE (Harold J.) (e) ANON	(a) Ool (b) Sh (c) HH (d) Exotic lep (e) Lep	Stevens 408 lots 24pp.	J; N

1903 – 1904	SOURCE	CONTENTS	AUCTIONEER & SALE CAT.	REFERENCE
October 20	(a) – (f) ANON	(a) – (b) Ool (c) Mam (d) Birds (e) Lib (f) Exotic lep	Stevens 332 lots 16pp.	J s.p.s.n.
November 19	BIDWELL (Edward), Part II	Ool	Stevens 240 lots 20pp.	J s.p.s.n.; N m.p.m.n.
November 24	(a) – (d) ANON (e) WHARTON (Dr. Henry Thornton) (f) – (k) ANON	(a) – (b) Ool (c) For sh (d) – (e) Birds (f) Min (g) HH (h) Exotic lep (i) HH (j) Australian & Pacific sh (k) Sh	Stevens 339 lots 20pp.	J; N; SAR
1904 January 12	BROWN (J. Allen), Ealing	Fos Min Nh	Stevens 174 + 22 bis lots 12pp.	**BMNHP**
January 26	(a) – (i) ANON	(a) Br lep (b) Exotic lep (c) S. & E.African lep (d) Lep (mimics) (e) Exotic lep (f) Exotic col Exotic lep (g) Lib (h) HH Mam (i) Birds	Stevens 359 lots 23pp.	J m.p.m.n.; SAR
February 1	(a) MACDONALD (K.C.) (b) ANON (c) NORWEGIAN EXPEDITION, 1902-1903 (d) – (e) ANON	(a) Indian ool (b) Ool (c) Birds & Ool from Novaya Zemla (d) – (e) Ool	Stevens 347 lots 26pp.	J s.p.s.n.; N
March 8	(a) WOODFORDE (F.C.) Market Drayton (b) ANON	(a) Br lep (b) Exotic lep Exotic col	Stevens 359 lots 21pp.	J m.p.m.n.; **CH** f.p.; **BMNHE**; SAR
March 29	(a) BUCKNILL (John.A.) (b) – (d) ANON (e) WHARTON (Dr. Henry Thornton) (f) ANON	(a) Br ool (b) Indian ool (c) Novaya Zemla ool (d) Ool (e) Birds (f) For lep	Stevens 451 lots 29pp.	J s.p.; N; JAG
April 26	(a) – (g) ANON	(a) Br lep (b) Exotic lep (c) Mam (d) HH Nh (e) Birds Nh (f) Ool Nh (g) W.Australian herb	Stevens 382 lots 18pp.	J; SAR
May 10	(a) NEVILL (Hugh) (b) – (f) ANON	(a) Sh (chiefly from Ceylon) (b) Mam (c) Fos Cor Madrapores Nh (d) Sh (e) Birds (f) Exotic lep Birds Nh	Stevens 390 lots 18pp.	**SPD**; J m.p.m.n. (of Nevill lots only); **NMW**
May 12	BIDWELL (Edward), Part III (final)	Ool Birds (young in down)	Stevens 390 lots 26pp.	J f.p.f.n.; N
May 17	MASON (Philip Brookes)	Lib (part I)	Stevens 330 lots 27pp.	J p.n.; **BMNHE**
May 18-19	NOBLE (Heatley)	Western Pal ool (part I)	Stevens 565 lots 43pp.	J p.n.; N f.p.
June 1	(a) – (e) ANON	(a) Exotic lep (b) Br lep (c) Exotic lep (d) Exotic lep Exotic col (e) Min	Stevens 352 lots 18pp.	J

1904 – 1905	SOURCE	CONTENTS	AUCTIONEER & SALE CAT.	REFERENCE
June 16	NOBLE (Heatley)	Western Pal ool (part II) Birds (young in down)	Stevens 314 lots 26pp.	**J** p.n.; **N** m.p. m.n.
June 21	MASON (Philip Brookes)	Lib (Part II)	Stevens 275 lots 28pp.	**J** p.n.; **BMNHE**
July 5	(a) – (h) ANON	(a) Indian lep (b) Br lep For lep (c) Br ool For ool (d) HH (e) N.Zealand birds (f) For birds (g) Sh (h) For ool	Stevens 466 lots 24pp.	**J** f.p.f.n.
July 12	MASON (Philip Brookes)	Lib (final part)	Stevens 287 lots 24pp.	**J** p.n.; **BMNHE**
July 13	COX (Dr. James C.) Sydney, New South Wales	Sh (part I)	Stevens 351 lots 16pp.	**NMW** p.n.
July 14	NOBLE (Heatley)	Western Pal ool (part III)	Stevens 271 lots 21pp.	**J** m.p.m.n.; **N** p.m.n.
August 23	(a) ALLMAN (George Johnston) (b) – (f) ANON	(a) Sh (b) Br lep Exotic lep (c) HH (d) Br birds (e) Indian ool (f) Br ool For ool	Stevens 381 lots 23pp.	**J** p.n. of (a); **NMW** s.p. (J.C. MELLVILL'S purchases)
September 27	(a) KING, natural history dealer, Great Portland Street (b) – (h) ANON	(a) Ins Sh Fish Nh (his stock of these groups) Lib (b) HH Mam (c) Min Crustacea (d) – (e) Br ool For ool (f) Br birds (g) Br ool (h) Ool	Stevens 518 lots 31pp.	**J**; **N**; **JAG**
October 18	(a) – (e) ANON	(a) Br birds For birds (b) Exotic birds Pal birds (c) N.American birds (d) Br ool For ool (e) Ool app Ornithological app	Stevens 287 lots 28pp.	**J**; **N**; **JAG**
October 25	(a) – (e) ANON (f) KNIGHT (J.W.)†	(a) New Guinea birds, 1882-84 (b) Sh (c) HH (d) Birds Ool (e) Exotic lep (f) Br lep For lep	Stevens 431 lots 23pp.	**J** p.n. (of Knight lots only)
December 8	(a) – (e) ANON (f) [WALLIS (S.H.)] (g) – (h) ANON	(a) – (c) Ool (d) St.Kilda ool (e) – (f) Ool (g) Ool (from P.NIELSEN of Eyrabakki (h) Birds	Stevens 284 lots 16pp.	**N**
December 20	(a) PINDER (Dr.)† (b) – (e) ANON	(a) Sh Fos Min Lib Nh (b) Lib (c) Bird skulls Animal skulls (d) Min (e) Exotic lep	Stevens 338 lots 20pp.	**J**
1905 January 17	(a) – (f) ANON (g) BIBBS (h) ANON	(a) Lib (b) Sh Fos Nh (c) Exotic lep (d) Ool (e) HH (f) Nh (g) Herb (h) Birds	Stevens 327 lots 20pp.	**J** s.p.s.n.; **BMNHE**
February 13	COX (Dr. James C.), Sydney, New South Wales	Sh (final part)	Stevens 249 lots 15pp.	**NMW**

1905	SOURCE	CONTENTS	AUCTIONEER & SALE CAT.	REFERENCE
February 14	(a) SMITHE (Rev. F.) (b) – (f) ANON	(a) Sh (b) Br ool (c) Br ool For ool (d) – (e) For ool (f) Peruvian lep	Stevens 521 lots 25pp.	J p.n. of (f); **N**; SAR
February 21	(a) URWICK (W.F.)† (b) SHIRLEY (S.E.), Ettington Park† (c) – (f) ANON	(a) Br birds (b) Dup birds (c) Birds (d) Exotic birds (e) Br ool (f) Ool	Stevens 452 + 17 bis (incl. 11 blank) lots 32pp.	J p.n. of (a); **N**; JAG
March 14-15	MASON (Philip Brookes), Part I	Br lep	Stevens 538 lots 34pp.	J p.n.; **CH** p.n.; **LC**; **BMNHE**; **GS**; SAR
March 16	(a) NOBLE (Heatley) (b) ANON	(a) Western Pal ool (final part) (b) Ool (Great Auk)	Stevens 346 lots 31pp.	**N** m.p.m.n.; J; JAG
March 28	(a) JONES (E.Harris), Putney (b) [ETHERIDGE] (c) WYNN (G.W.), Hampton-in-Arden, Warwickshire (d) ANON	(a) Br macrolep Br microlep (b) Br lep (c) Br lep (d) Exotic lep	Stevens 439 lots 28pp.	J m.p.m.n.; SAR
April 14	HAWKINS (John)†	Min	Foster 296 lots 11pp.	**BMNHM** m.p. m.n.
April 18	(a) – (e) ANON	(a) Min Sh (b) HH Sh (c) Birds Exotic lep (d) – (e) Ool	Stevens 418 + 29 bis lots 26pp.	J f.p.f.n.; **N**; JAG
May 16-17	MASON (Philip Brookes), Part II	Br lep	Stevens 562 lots 36pp.	J p.n.; **GS**; **CH**; **BMNHE**; **LC**; SAR
May 23	(a) – (b) ANON	(a) Br lep (b) Exotic lep	Stevens 414 lots 24pp.	J s.p.s.n.; **BMNHE**
June 6-7	(a) ANON (b) IRVING (c) – (e) ANON (f) [GODDARD] (g) ANON (h) CAMPBELL (Sir John W.P.), Orde of Kilmorey (i) – (j) ANON	(a) – (b) Min (c) Sh (d) Egypt & Uganda hh (e) Exotic lep (f) Br lep (g) Exotic lep (h) Birds (i) Birds (j) Br ool For ool	Stevens 678 lots 39pp.	J p.n. (of lep only); SAR; JAG
June 21	(a) ANON (b) CAMPBELL (Sir John W.P.), Orde of Kilmorey	(a) Lib (b) Lib	Stevens 196 lots 22pp.	J s.p.s.n.
July 11	(a) ANON (b) GOODALL (Dr.); ARGENT; EVANS; JOHNSON (c) – (g) ANON	(a) Sh (b) Sh (c) Fos Min (d) Great Tibetan Stag, Tibetan Bear, Clouded Leopard (all from Lhasa) (e) *Aepyornis* (f) HH (g) Exotic lep	Stevens 465 lots 24pp.	J; NMW; SAR
August 22	(a) BARTLETT (A.D.), late of the Zoological Gardens, Regents Park† (b) – (c) ANON	(a) 1034 autograph letters mostly to the same person [A.D. BARTLETT] (b) Lib Ornithological plates (c) Exotic lep	Stevens 248 lots 16pp.	J; SAR
October 10	(a) [HAWKINS (John)] (b) ANON	(a) Min (b) Br lep Exotic lep Fos HH	Stevens	*Nature*, April 1905:ccxxiv

1905–1906	SOURCE	CONTENTS	AUCTIONEER & SALE CAT.	REFERENCE
October 10	(a) – (d) ANON	(a) Birds (b) – (c) Exotic lep (d) For lep (incl. Chinese lep)	Stevens 353 lots 21pp.	J s.p.s.n.; SAR
October 11	BUNYARD (P.F.)	Ool	Stevens 204 lots 16pp.	N; JAG
October 24	DAY (George O.), Knutsford	Br lep	Stevens 360 lots 24pp.	J p.n.; CH m.p. m.n.; BMNHE; LC; SAR
November 7	(a) BEAUMONT (A.)† (b) BARRETT (C.G.)† (c) ANON	(a) Lib (b) Lib (c) Lib	Stevens 415 lots 38pp.	J p.n.; CH; BMNHE
November 14	(a) BRAUER (Richard), Knutsford† (b) – (c) ANON	(a) Exotic lep (b) Ins (c) HH	Stevens 352 lots 21pp.	J m.p.m.n.; BMNHE
November 28-29	MASON (Philip Brookes), Part III (final)	Br lep	Stevens 526 lots 40pp.	CH s.p.s.n.; J p.n.; LC; BMNHE; H; SAR
December 5	(a) WALLIS (S.) (b) – (e) ANON	(a) – (b) Ool (c) – (d) Birds (e) Ool	Stevens 365 lots 25pp.	J f.p.f.n.; N; JAG
December 19	(a) – (d) ANON	(a) Exotic lep (b) Birds Mam (c) Min (d) Lib	Stevens 305 lots 16pp.	J; SAR
1906 January 17	(a) PIDSLEY (Helman)†[1] (b) ANON	(a) Ool (b) Ool (Great Auk)	Stevens 280 lots 32pp.	J f.p.f.n.; N
February 5	(a) BEAUMONT (A.)† (b) ANON	(a) Br macrolep Br microlep Col Hym Dipt Neur Orth (b) Lep	Stevens 344 lots 26pp.	CH; BMNHE; J s.p.s.n.; H; SAR
February 20	BAZETT (Mrs.), Reading	Br lep	Stevens 361 + 3 bis lots 20pp.	J p.n.; CH; BMNHE
March 13	BARRETT (C.G.)†	Br lep	Stevens 364 lots 24pp.	J p.n.; CH m.p.; LC; BMNHE; SAR; H
March 17	CORBETT (John), M.P., Impney near Droitwich†	Min (table with labradorite felspar top)[2]	– house sale	Birmingham Pos 17. iii. 1906
March 26	(a) Exotic lep (b)	(a) Exotic lep (b) Lep Nh (c) Butts (from Columbia) Col (from Columbia) (d) HH Nh (e) S.African mam (f) Br birds (g) For birds Lep Hym	Stevens 398 lots 21pp.	LC; J; SAR

[1] Pidsley of Broadclyst – author of *Birds of Devon* (1890).

[2] 'At Oxford yesterday I had a good look at a magnificent table-top of polished labradorite and microcline in Prof. Vincent's room in the Geology Dept. Made by Bryce Wright before 1890, a description is given in the *Pall Mall Gazette*, 8. i. 1890' (P. Embrey *in litt.* 1. xi. 1974).

1906 – 1907	SOURCE	CONTENTS	AUCTIONEER & SALE CAT.	REFERENCE
March 27	(a) BARRETT (C.G.)† (b) URWICK (W.F.)† (c) ANON	(a) Br microlep European lep S.African lep Nh (b) Br lep (c) Br lep	Stevens 313 lots 26pp.	J p.n.; CH m.p.; LC; BMNHE; H
April 25	(a) PIDSLEY (Helman)† (b) WRIGHT (W.C.), Belfast (c) ANON	(a) Ool (final part) (b) – (c) Ool	Stevens 430 lots 36pp.	BMNHZ; J s.p. s.n.[1]; JAG
May 1	(a) DOBRÉE-FOX (Rev.) (b) ANON	(a) Br lep (b) Lep Nh	Stevens 259 lots 20pp.	J m.p.m.n.; CH; BMNHE
May 2	ANON	Exotic lep	Stevens 352 lots 20pp.	J m.p.m.n.
May 23	(a) DALE (C.W.)† (b) PIDSLEY (W.E. Helman)† (c) BARRETT (C.G.)† (d) BEAUMONT (A.)	(a) Lib (b) Ornithological lib (c) Lib (d) Lib	Stevens 393 lots 47pp.	J m.p.m.n.; BMNHE
May 29	(a) GREENE (Rev. Joseph)†[2] (b) HELPS (J.A.)†	(a) Br lep (b) Br lep	Stevens 395 lots 30pp.	J m.p.m.n.; CH m.p.; LC; BMNHE
May 30	LINGWOOD (Henry), The Chestnuts, Needham Market, Suffolk	Br lep	Garrod Turner, *house sale*	*Nature*, May **1906**:xxxi
June 19-20	(a) SEQUEIRA (Dr. James Scott) (b) ANON (c) – (d) [WILDE (J.P.), Birmingham; et al]	(a) Br lep (part coll.) (b) Br lep Exotic lep (c) HH Sh Nh (d) Min Sh Ool Birds	Stevens 600 lots 42pp.	BMNHE; J; SAR
October 2	(a) MORRIS (C.H.), Lewes (b) – (c) ANON	(a) Br lep (b) Exotic lep (c) Exotic col	Stevens 382 + 5 bis lots 19pp.	J m.p.m.n.; BMNHE; CH; SAR
October 30	(a) PIDSLEY (W.E. Helman)† (b) BIDWELL (E.) (c) – (f) ANON	(a) Br birds For birds Ool (b) HH (c) – (f) Ool	Stevens 510 lots 31pp.	J p.m.n.; BMNHZ; N m.p.f.n.; JAG
November 20	MATHEW (Gervase F.)	Br lep	Stevens 524 + 4 bis lots 28pp.	J p.n.; CH f.p.; BMNHE; SAR
November 27	(a) – (h) ANON	(a) Exotic lep (b) Exotic col Exotic neur (c) Exotic lep (d) HH Rept (e) Fos Min (f) Birds (g) Ool (h) Sh	Stevens 530 lots 31pp.	J s.p.s.n.; SAR
December 18	RAMSDEN (Hildebrand)	Br lep For lep Lib	Stevens 359 lots 28pp.	J p.n.; BMNHE
1907 January 21-22	(a) ANON, incl. BERTLING & BARTLETT (b) – (i) ANON	(a) Ool (b) Birds (c) For lep (d) Exotic lep (e) Br lep (f) Lep Col Exotic hym Exotic hem (g) Sh Min HH Nh (h) HH (i) Sh	Stevens 778 lots 39pp.	J s.p.s.n.; SAR; JAG

[1] In a MS note in the Janson cat., O.E.Janson wrote: 'Most of the Cuckoo's eggs in this sale are a *fraud* put in by H.Williamson, 10 Watling St., Manchester *teste* Bidwell, Marsden, Massey et al'.

[2] Greene (1824–1906) was the author of *The Insect Hunters' Companion* (1863) comprising his celebrated *Essay on Pupa Digging* first published in the *Zoologist* for 1857.

1907	SOURCE	CONTENTS	AUCTIONEER & SALE CAT.	REFERENCE
March 19	LANG (Rev. H.C.)	Pal butts	Stevens 327 lots 27pp.	J p.n.; BMNHE SAR
April 9	(a) – (c) ANON (d) WILSON (James) (e) ANON (f) ANON incl. Dr. BJORKBORN, Pitca, Lapland (g) – (h) ANON	(a) Exotic lep (b) Sh Min Birds (c) Birds (d) HH (e) Exotic lep (f) Birds (g) Ool (h) Birds	Stevens 418 lots 28pp.	J f.p.f.n.; N; SAR; JAG
May 15	(a) ROBSON (J.E.) Hartlepool† (b) ANON (c) URWICK (R.H.) (d) ALLEN (L.F.)	(a) Br macrolep Br microlep (b) Lib (c) – (d) Br lep	Stevens 344 + 14 bis lots 28pp.	J p.n.; CH f.p. f.n.; BMNHE; SAR
May 28-29	(a) – (e) ANON (f) EMSLEY (F.) (g) KITCHEN (V.P.) (h) BECK-FORD (William), Fonthill Abbey, Wiltshire†	(a) Ool (b) Sh Mam Nh (c) Ool (d) HH Rept Birds Min Ins Nh (Guatemala) Exotic lep (e) Sh (f) Br lep (g) Br lep (h) Sh	Stevens 604 lots 40pp.	J; N; JAG
September 23-24	(a) – (k) ANON	(a) Exotic col (b) Exotic lep (c) Aberdeen lep (d) Br lep (e) African hh (f) Sh Cor (g) Br sh (h) Exotic sh (i) Exotic birds (j) Br birds (k) Dup ool	Stevens 648 + 24 bis lots 40pp.	J s.p.s.n.; JAG
October 21	DA COSTA (S.J.)†	Sh Lib	Stevens 438 + 5 bis lots 28pp.	J p.n.; SPD; NMW; CH
October 22	RAYNOR (Rev. G.H.)	Br lep (part I)	Stevens 350 lots 24pp.	J p.n.; CH p.m. BMNHE; SAR
October 28	STOATE (William)	Ool	Stevens 294 lots 23pp.	J; SAR
November 4	(a) [WALLIS (S.H.), Weymouth] (b) – (h) ANON	(a) Ool (b) Nh (c) Exotic lep Br lep (d) Malay lep (e) Col (f) Aberdeen lep (g) HH (h) Birds	Stevens 403 + 34 bis lots 23pp.	J s.p.s.n.; CH; BMNHE; N; SAR
November 5	(a) RAYNOR (Rev. G.H.) (b) – (f) ANON	(a) Br lep (part II & final) (b) Br macrolep Br microlep (c) Br lep Ool (d) – (e) Exotic lep (f) Exotic lep Ins Nh	Stevens 381 lots 29pp.	J p.n.; CH m.p. m.n.; BMNHE; SAR
November 14	DAVIS (J.W.)†	Fos (fish) Fos (rept) Min Fish Herb	Hepper, *Leeds* 280 lots 11pp.	BMNHP
November 19	(a) MOORE (F.)† (b) ANON (c) ASH (Rev. C.D.)† (d) ANON	(a) Lib Exotic lep Exotic ins Cab (b) Lib (c) Br lep (d) Lep Col Fos Sh Nh Cab	Stevens 383 + 9 bis lots 32pp.	J p.n.; CH; BMNHE; SAR
December 2	BACKHOUSE (James), York†	Fos (mam) Great Auk (skeleton)	Stevens 293 lots 18pp.	BMNHP

1907–1908	SOURCE	CONTENTS	AUCTIONEER & SALE CAT.	REFERENCE
December 3	(a) BARRETT (C.G.)† (b) CROSS (W.J.), Ely[1]	(a) Microlep (b) Macrolep Microlep	Stevens 298 lots 30pp.	J p.n.; CH s.p. s.n.; BMNHE; SAR
December 9-10	RUSTAFJAELL (Robert de), F.R.G.S.	Fos Min, 14	Stevens 245 lots 18 lots + 8 plates	BMNHP
1908 January 14	(a) AULD (H.A.) (b) ANON (c) SHEPHERD (A.H.)† (d) – (e) ANON	(a) – (c) Br lep (d) Exotic lep (e) Sh	Stevens 292 lots 23pp.	J p.n.; BMNHE; SAR
January 21	(a) – (g) ANON	(a) Min (b) Sh (incl. some from Prof. STEWART coll.) (c) Exotic lep (d) Aberdeen lep (e) Queensland nh (f) Exotic col (g) Birds HH	Stevens 447 lots 25pp.	J s.p.s.n.; BMNHE; SAR inc.
January 30	MIDDLEBROOK MUSEUM[2]	Ool (incl. Great Auk) Nh	Debenham & Storr	Parkin (1911:23)
February 11	(a) STEWART (Prof.) (b) ANON	(a) – (b) Lib	Stevens 288 lots 30pp.	J m.p.m.n.
February 25	(a) – (f) ANON	(a) Exotic birds (b) Ool (c) Sh (d) Lep (C.G.BARRETT, G.O. DAY, A. BEAUMONT & P.B. MASON dups) (e) Exotic lep (f) Lep Exotic col	Stevens 413 lots 28pp.	J m.p.m.n.; SAR; JAG
March 3	TUNSTALL (W.)	Br lep	Stevens 337 lots 24pp.	J p.n.; CH p.m.n.; BMNHE; SAR
March 17	(a) GARDNER (P.T.), Cambridge (b) – (e) ANON	(a) – (b) Br lep (c) Lep (d) App (e) Exotic butts Ins	Stevens 283 lots 23pp.	J m.p.m.n.; CH; BMNHE; SAR
March 18	(a) STOATE (William) (b) ANON	(a) Ool (final part) (b) Br ool For ool	Stevens 314 lots 23pp.	J; N
April 28	TURLE (W.H.)	Br ool For ool	Stevens 288 lots 23pp.	J; N p.s.n. JAG
May 19-20	(a) – (b) ANON (c) BOLTON (Gambier) (d) MILLAIS (J.G.) (e) STURT (Fred)† (f) ANON (g) LILFORD (h) – (i) ANON (j) GIBBS	(a) Exotic birds (b) Fos Sh Min (c) HH (d) Br birds (albinos) (e) Birds (f) Ool (g) Ool (h) Sh (i) Br lep For lep Nh (j) Br lep	Stevens 622 lots 38pp.	J m.p.m.n., BMNHZ; SAR; JAG
June 2	GOSS (Herbert)†	Br lep	Stevens 276 lots 19pp.	J p.n.; CH s.p. s.n.; BMNHE

[1] A note in the cat. states that the 'diaries relating to the collection can be seen'.

[2] This was T.B.Middlebrook's Free Museum in the 'Edinburgh Castle' public house, Mornington Road, Regents Park, London N.W.

1908–1909	SOURCE	CONTENTS	AUCTIONEER & SALE CAT.	REFERENCE
July 14-15	(a) HARRISON (J.), Barnsley (b) JOBSON (H.W.) (c) HEATH (Dr. A.E.) (d) – (e) ANON (f) THATCHER (Mrs.) (g) ANON	(a) – (b) Br lep (c) Br lep Exotic lep Exotic col (d) Lib (e) German New Guinea lep (f) Exotic butts (g) Ool Nh	Stevens 653 + 11 bis lots 42pp	J p.n.; CH s.p.; BMNHE; SAR
October 6	(a) – (g) ANON	(a) Ool (b) Birds Nh (c) Ool Birds (d) Ool (e) Ool (ex LILFORD coll.) (f) Birds Mam Nh (g) HH Mam Fos Nh	Stevens 378 + 37 bis lots 23pp.	N f.p.; JAG
October 27	(a) THORNTHWAITE (W.H.E.)† (b) – (c) ANON (d) RENTON (W.) (e) ANON	(a) Br macrolep (b) Ins (c) Br lep (d) Br lep (e) Exotic butts	Stevens 388 + 4 bis lots 28pp.	CH m.p.m.n.; J m.p.m.n., SAR; BMNHE
1909 January 26-27	(a) SCHILL (C.H.), incl. J. CHAPPELL† et al. (b) SMALLPIECE (A.M.), incl. A.E. CANNON, Aberdeen (part coll.) (c) SCHILL (C.H.) (d) – (e) ANON	(a) Br macrolep Br microlep (b) Br lep (c) Col (World Carabidae) (d) Exotic butts (e) Lep	Stevens 682 + 11 bis lots 39pp.	J m.p.m.n.; CH f.p.; BMNHE
February 9	(a) ANON (b) ATHERTON (J.) (c) MILLS (Dr. Y.H.), Haverfordwest (d) – (f) ANON (g) KENT (W. Saville)† (h) USSEL (Madame)	(a) – (b) Ool (c) Ool (Cuckoos with fosterers) (d) Ool (e) Birds HH (f) Sh (g) Cor Birds Sponges (h) Ool (Great Auk)	Stevens 359 + 41 bis lots 22pp.	J f.p.; BMNHZ; N f.p.
February 23-24	MADDISON (T.)†, Part I	Br lep	Stevens 600 lots 38pp.	J p.n.; CH m.p. BMNHE; SAR
March 9-10	(a) MADDISON (T)†, Part II (final) (b) SCHILL (C.H.) (c) – (d) ANON (e) THORNTHWAITE (W.H.E.)† (f) KENT (W. Saville)† (g) – (j) ANON	(a) Br lep (b) – (g) Lib (h) Br lep (i) Ins (j) Exotic birds Exotic lep Nh	Stevens 619 + 8 bis lots 36pp.	J m.p.m.n.; CH; BMNHE
March 23	PARDOE (J.)†	Br lep	Stevens 299 lots 15pp.	CH s.p.s.n.; J s.p.s.n.; BMNHE; SAR inc.
April 6	(a) – (b) ANON (c) CROMBIE (Dr.)† (d) ANON (e) JACKSON (Mrs.) (f) – (i) ANON	(a) Lep (incl. many French microlep) Col Hym Ins Gall-flies with galls (b) Fish (c) – (d) Ool (e) E.African ool (f) Nh (g) HH (h) Sh Eocene fos (i) Lib	Stevens 512 lots 24pp.	J s.p.s.n.; BMNHE; SAR JAG
May 11-12	(a) SCHILL (C.H.) (b) PROUT (Louis B.) (c) HORLEY (Dr. W.)	(a) World col Br col (b) Br macrolep (excluding Geometridae) Br microlep (c) Br lep	Stevens 693 lots 44pp.	J s.p.s.n.; CH p.n. of (b); BMNHE; SAR
May 25	(a) – (c) ANON	(a) Sh (b) Exotic lep (c) Birds HH	Stevens 394 lots 22pp.	J; N; AB
August 17	(a) 'Contents of a small Natural History Museum' (b) ANON (c) HORLEY (W.L.) (d) – (f) ANON	(a) Birds Ool Sh Nh (b) Ool (c) Birds (d) HH Mam (e) Min Sh (f) Exotic lep Exotic col Nh	Stevens 310 lots 15pp.	J; N
September 21	(a) – (b) ANON (c) [BOOTH (H.T.)]	(a) Lep Col Ins (b) Exotic lep (c) Ool	Stevens 407 lots 20pp.	BMNHZ; J p.n. (of lots 1–100 only); N s.p.f.n.

1909–1910	SOURCE	CONTENTS	AUCTIONEER & SALE CAT.	REFERENCE
October 5-6	(a) GOLDTHWAIT (O.C.) (b) PARTRIDGE (Lt. Col.Charles E.) (c) CLAXTON (Rev.William) (d) LINNELL (John), Redhill, Surrey† (e) PRESTON (Rev.) (f) ANON	(a) – (c) Br lep (d) Br col (with ms. diaries) (e) Brazilian ins (f) Exotic lep Exotic col.	Stevens 679 + 10 bis lots 42pp.	J m.p.m.n.; CH f.p.; BMNHE
October 19	(a) LINTER (Miss J.E.)† (b) ANON (c) ANON incl. BARTLETT & BERTLING	(a) Sh (b) – (c) Ool	Stevens 323 + 8 bis lots 17pp.	NMW; J
October 29	DAMON (R.F.), Weymouth, Part I	Fos Min	Foster 308 lots 12pp.	J f.p.; BMNHM; BMNHP
November 2-3	CLARK (J.A.)†	Br lep (part I)	Stevens 448 + 49 bis lots 27pp.	J p.n.; CHm.p.; LC; BMNHE; MAR n. (of butts only)
November 16	(a) BARKER (H.W.)† (b) PALLISER (H.G.) (c) ANON (d) BERTLING (A.) incl. BARTLETT	(a) Br lep Lib (b) Exotic lep (c) Entomological lib (d) Ool	Stevens 376 lots 25pp.	J m.p.m.n.; CH f.p.f.n.; BMNHE; BMNHZ; SAR
December 7-8	CLARK (J.A.)†	Br lep (part II)	Stevens 601 lots 34pp.	J p.n.; CH p.m.n; LC m.p.; BMNHE; SAR
1910				
January 15	ANON	Ool Ins Nh Irr	Stevens 317 lots 22pp.	BMETH
January 18-19	(a) CLARK (J.A.)† (b) – (c) ANON (d) HORLEY (W.L.) (e) – (f) ANON	(a) Exotic lep (b) Sh Min Br birds Br ool (c) Birds Ool (d) Birds (e) Mam (f) Sh	Stevens 568 lots 23pp	J m.p.m.n.; BMNHE; BMNHZ; N; JAG
February 8	(a) – (c) ANON (d) WILLMOTT (Collis) (e) – (f) ANON	(a) Exotic lep Exotic col Exotic ins (b) Min (c) Sh (d) Birds (e) Ool (*Aepyornis maximus*) (f) Ool	Stevens 451 lots 19pp.	J s.p.s.n.; BMNHZ; N m.p.f.n.; JAG
February 22-23	(a) CLARK (J.A.)† (b) ANON (c) WOODFORD (C.M.) (d) ANON	(a) Br macrolep Br microlep (final part) (b) Exotic ins (c) Solomon Is. lep (d) Br lep Br col.	Stevens 635 + 3 bis lots 39pp.	J p.n.; CH f.p. f.n.; BMNHE; SAR
March 8	(a) BROCKHOLES (James) Claughton† (b) – (e) ANON (f) [? SKINNER (K.L.)]	(a) Ool (b) Birds (c) Min (incl. some from Broken Hill Mines, New South Wales) (d) Min (incl. native copper & Cornwall malachite (e) Br lep Exotic lep (f) Ool	Stevens 489 lots 20pp.	BMNHE; BMNHZ; J; N m.p.s.n. (of ool only); SAR; JAG
April 12-13	(a) – (d) ANON (e) BLAIR (Alexander), Crieff (f) – (g) ANON	(a) Br lep (b) Exotic lep (c) Lib (d) – (e) Ool (f) Fos Min (g) Birds	Stevens 637 + 53 bis lots 32pp.	J f.p.f.n.; JAG
April 26	(a) MORRIS (Rev. F.O.)† (b) OLIVER (G.B.) (c) McARTHUR (H.)† (d) GROSE-SMITH (H.)	(a) Br ins For ins (b) Br lep (c) Br lep (d) Exotic lep (dups)	Stevens 445 lots 23pp.	J p.n.; CH f.p.; BMNHE

1910–1911	SOURCE	CONTENTS	AUCTIONEER & SALE CAT.	REFERENCE
May 24	(a) FISHER (Rev. F.H.)† (b) – (c) ANON (d) GROSE-SMITH (H.)	(a) Br lep (b) Br lep (c) Exotic lep (d) Exotic lep	Stevens 322 + 31 bis lots 19pp.	J m.p.m.n.; **CH** s.p.s.n.; **BMNHE**
June 7	(a) – (b) ANON (c) BULLEN (Rev. R. Ashington) (d) – (i) ANON (j) SKINNER (E.R.) (k) GREEN (Joseph F.)	(a) Lep Col App (b) – (e) Sh (f) Min Fos (g) HH (h) Ool (Great Auk) (i) Ool (*Aepyornis maximus*) (j) Ool (k) Ool (Cuckoo)	Stevens 462 lots 25pp.	J m.p.m.n.; **NMW; BMNHZ N**
June 8	DAMON (Robert F.) Weymouth, Part II (final)	Sh Fish Rept.	Stevens 351 lots 18pp.	**NMW; J**
July 5	(a) SAUNDERS (G.S.)† (b) SAUNDERS (Edward)† (c) JACOBY (Martin)†	(a) – (c) Lib	Stevens 327 lots 27pp.	J p.n.
September 27	(a) ARMITAGE (E.), R.A.† (b) ANON	(a) European col Exotic col (b) Exotic lep	Stevens 407 lots 23pp.	J m.p.m.n.; **SAR**
October 25-26	(a) RAYNOR (Rev. G.H.) (b) TURNER (E.) (c) ANON	(a) Lep (*Abraxas grossulariata*) (b) Br lep (c) Ins Exotic lep	Stevens 888 + 4 bis lots 40pp.	J m.p.m.n.; **CH** f.p.f.n.; **BMNHE; SAR**
November 8	(a) – (c) ANON	(a) Ool (b) Ool (Cuckoo) (c) Ool	Stevens 395 lots 20pp.	J m.p.; **N**
December 6-7	(a) – (b) ANON (c) RATTRAY (Colonel) (d) – (e) ANON (f) NECK (J. Frederick)†	(a) Exotic lep (b) Birds Ool (c) Indian ool (d) HH (e) Sh (f) Min	Stevens 776 lots 32pp.	J s.p.; **BMNHE; BMNHZ; N** f.p. f.n.; **SAR; JAG**
1911 January 10	(a) BARCLAY (Colonel Hanbury)† (b) ANON (c) RATTRAY (Colonel)	(a) Birds (b) Ool (c) Indian ool	Stevens 400 lots 19pp.	**BMNHZ; J; N** m.p.s.n.
January 17	JOICEY (J.J.)	Dup exotic lep	Stevens 400 lots 15pp.	J f.p.; **BMNHE**
January 31	BARCLAY (Colonel Hanbury)†	Ool (part I)	Stevens 280 lots 24pp.	J m.p.m.n.; **BMNHZ; N** p.m.n.; **JAG**
February 7	BARCLAY (Colonel Hanbury)†	Lib	Stevens 365 lots 31pp.	J m.p.m.n.; **BMNHZ; N** f.p.
February 21	(a) – (e) ANON	(a) Ool (b) HH Mam (c) South American butts (d) Exotic lep (e) Min	Stevens 387 lots 15pp.	**BMNHZ; N; SAR**
March 28	BARCLAY (Colonel Hanbury)†	Ool (final part)	Stevens 371 lots 22pp.	**BMNHZ; N; JAG**
April 11	(a) TUTT (J.W.)† (b) – (c) ANON (d) COX (C. Stanley Bell) (e) SIMPSON (Thomas) Ealing†	(a) Lep (part I) (b) Exotic lep (c) Sh Cor Fos (d) Sh Lib (e) Lib	Stevens 494 + 9 bis lots 25pp.	**CH** s.p.s.n.; J s.p.s.n.; **LC** m.p. **BMNHE** p.n.; **SAR**

1911–1912	SOURCE	CONTENTS	AUCTIONEER & SALE CAT.	REFERENCE
May 9-10	CLARK (J.A.)	Lib, 99	Sotheby 896 lots 67pp.	J p.m.n.
May 23-24	(a) – (b) ANON (c) SIMPSON (Thomas), Ealing† (d) – (e) ANON	(a) Pal lep Exotic lep Exotic col (b) European butts (c) Birds Ool (d) Mam HH (e) Sh Min Fos Nh	Stevens 729 lots 34pp.	J p.n.; BMNHE; BMNHZ; N
June 13-14	ARNOLD (George), Milton Hall, Gravesend, Part I	13 vi: Nh + much Irr 14 vi: Fos Min Sh Nh, 79	Stevens, *house sale* 900 lots 48pp.	BMETH
July 18	(a) REED (J.T.T.)† (b) – (d) ANON (e) [? MURRAY (A.)] (f) – (h) ANON	(a) – (b) Ool (c) Min Nh (d) Lib (e) Br lep (f) Exotic lep (g) Butts (h) HH	Stevens 436 lots 22pp.	J s.p.s.n.; BMNHE; N
September 19	(a) TUTT (J.W.)† (b) – (d) ANON	(a) Lep (part II) (b) Br lep For lep (c) HH (d) Nh	Stevens 278 lots 15pp.	J m.p.m.n.; LC m.p.; CH; BMNHE
October 24	(a) ANON (b) RATTRAY (Colonel) (c) – (d) ANON	(a) Ool (b) Indian ool (c) Ool (d) HH	Stevens 436 lots 19pp.	J s.p.s.n.; BMNHZ; SAR
November 7	(a) COTTAM (A.) (b) – (e) ANON	(a) Br lep (b) Lep (c) Lib (d) Lep (e) Min	Stevens 405 lots 23pp.	J m.p.m.n.; CH LC s.p.f.n.; BMNHE; SAR
December 5	(a) – (d) ANON	(a) Ool (b) Br birds For birds (c) Sh (d) HH	Stevens 343 lots 16pp.	J f.p.f.n.; BMNHZ
December 19	(a) TUTT (J.W.)† (b) – (c) ANON (d) 'A Collector's Duplicates' (e) ANON	(a) Pal butts (b) Br lep (c) Br lep (d) Exotic butts (e) Exotic dipt	Stevens 410 lots 24pp.	J m.p.m.n.; CH s.p.; BMNHE
1912				
January 9	(a) SCHILL (C.H.) (b) NEVINSON (Basil George)† (c) WOLLASTON (T. Vernon)†	(a) Lib (final part) (b) – (c) Lib	Stevens 355 lots 27pp.	J p.n.; CH f.p. f.n.; BMNHE
February 6	(a) MALING (W.) Newcastle-upon-Tyne† (b) ANON (c) RIDING (W.S.)† (d) ANON	(a) Br macrolep Br microlep (b) Lep (c) Br lep (d) Lib	Stevens 381 + 13 bis lots 26pp.	J m.p.m.n.; LC m.p.s.n.; BMNHE; CH
February 20	(a) – (b) ANON	Ool	Stevens 373 lots 18pp.	J f.p.f.n.; BMNHZ; N
February 23	CAIRNS (Robert), Hurst, Ashton-under-Lyne†	Sh Lib	Capes, Dunn, *Manchester* 274 lots 20pp.	NMW p.m.n.
March 5	(a) RAYNOR (Rev. G. H.) (b) [WEBSTER (Dr.), Brighton] (c) ANON	(a) Br butts Lep (*Abraxas grossulariata*) (b) Br lep (lots 180–322) (c) Exotic lep	Stevens 399 + 5 bis lots 20pp.	J m.p.m.n.; CH s.p.; LC s.p.s.n.; BMNHE
March 18-21	ANON	Mam HH	Hudson's Bay Company 2702 lots 124pp.	BMNHZ

1912	SOURCE	CONTENTS	AUCTIONEER & SALE CAT.	REFERENCE
March 19	(a) ANON (b) [MARSDEN (H.W.)] (c) – (i) ANON	(a) -- (b) Ool (c) Lib (d) Sh (e) Col (f) App (g) Nh (h) Br lep Exotic lep (i) HH	Stevens 451 lots 25pp.	BMNHE; CH; J N m.p.s.n.
April 17	(a) WALTER (H.F.), Papplewick[1] (b) [COPELAND (A.T.), Shifnel] (c) SMYTH (Lady Greville), Ashton Court, Somerset (d) BIDWELL (E.) (e) NEWCOMBE (S.Prout) (f) ANON	(a) – (b) Ool (c) Ool (Great Auk, 2) (d) Ool (e) Min (f) Min	Stevens 440 lots 23pp.	J f.p.; N m.p. s.n.
April 23	(a) CASSAL (Dr.), Ballaugh, Isle of Man† (b) ANON (c) TUTT (J.W.)† (d) – (e) ANON	(a) Br macrolep Br microlep Ins (incl. Hym Dipt Trichoptera Col Neur) (b) Lib (c) Br macrolep Br microlep (penultimate part) (d) Lep (e) Exotic lep (incl. Dutch New Guinea lep collected by C. & F. PRATT)	Stevens 428 + 26 bis lots 23pp.	J m.p.m.n.; CH s.p.; BMNHE; LC
May 6	(a) – (g) ANON	(a) Br lep Exotic lep (b) Br lep (c) Exotic lep (d) Ool (e) Birds (f) Sh Min (g) HH	Stevens 483 lots 30pp.	J s.p.s.n.; BMNHE; N
May 22	(a) ANON (b) [MARSDEN (H.W.)] (c) – (d) ANON	(a) Ins (incl. Col Exotic lep) (b) Ool (c) HH (d) Br sh	Stevens 483 lots 22pp.	J s.p.s.n.; BMNHE; N
July 9	(a) – (e) ANON (f) HARRISON (A), F.E.S. (g) ARNOLD (George), Gravesend	(a) Br lep (b) Exotic lep (c) Sh Min (d) Birds (e) HH (f) Lib (g) Fos Min Sh Nh	Stevens 386 lots 22pp.	J m.p.m.n.
September 24	(a) TUTT (J.W.)† (b) ADAMS (H.J.)† (c) ROGERS (E.A.) (d) CONQUEST (G.H.), Westcliff-on-Sea†	(a) Lep (final part) (b) Br ins (incl. Col Dipt Hym Hem Neur) Exotic col (c) Br lep (d) Br macrolep Br microlep	Stevens 451 lots 27pp.	J p.n.; CH m.p. m.n.; BMNHE; LC f.p.f.n.
October 8	(a) WHYMPER (Edward)† (b) KING (Henry A.)† (c) – (d) ANON (e) STEPHENS (Alfred) (f) ANON	(a) For ins (incl. Lep Neur Col Dipt Orth Hym Hem) Nh (b) Br lep (chiefly from Harwich & Ipswich districts) (c) Sh Fos (d) Min Fos (e) Lep (dups) (f) HH Ins Nh	Stevens 446 + 10 bis lots 25pp.	J m.p.m.n.; CH BMNHE
October 22	ANON	Ool	Stevens 370 lots 20pp.	J; N
November 12-13	(a) HODGES (A.J.) (b) HARRISON (A.)†; MAIN (Hugh) (c) [BOOTH (Dr.)] (d) [TAYLOR (W.G.)] (e) [TEMPANY (Mrs.)]	(a) – (b) Br lep (c) – (e) Exotic lep	Stevens 597 lots 35pp.	J m.p.m.n.; CH BMNHE
November 21	(a) PROCTOR (Major F.W.), Part I (b) ANON	(a) Western Pal ool (b) Ool (Great Auk)	Stevens 289 lots 25pp.	J; BMNHZ; N p.m.n.

[1] A MS. note by O.W.Tancock in the Nichol's cat. reads: 'H.F. Walter bought the collection in 1851. He was a great friend of Wolley and had many of his eggs. He left his increased collection to his son John Henry Walter who carried the eggs and cabinets to a new house at Drayton near Norwich — placed them in a damp room.........' 'Many of the eggs sold were not in good condition...........'

1912–1914	SOURCE	CONTENTS	AUCTIONEER & SALE CAT.	REFERENCE
December 10	(a) JEFFREY (W. Rickman), Ashford, Kent (b) BAXTER (T.) (c) HARWOOD (W.H.) (d) – (e) ANON	(a) – (b) Br lep (c) Lep (*Abraxas grossulariata*) (d) Exotic ins (e) HH	Stevens 411 lots 23pp.	**CH** m.p.m.n.; **GS**; **J** m.p.m.n.; **BMNHE**
1913 January 20-24	DE RUSTAFJAELL (Robert)	22 i: Egyptian mummified mam and birds, 3	Sotheby 1051 lots 84pp.	**BML** p.n.; **S** p.n. copy
February 18	(a) SEQUEIRA (Dr. J.S.)† (b) BUCKELL (E.) (c) HENSLOW (Rev. Prof.)† (d) ANON	(a) Br macrolep Br microlep (b) Br lep (c) Br ins (incl. Hym Lep Col Orth) (d) Exotic lep	Stevens 525 lots 31pp.	**J** m.p.m.n.; **LC** f.p.; **CH**; **BMNHE**; **SAR**
February 27	BULOW (Carl)	Sh	Stevens 448 lots 24pp.	**NMW** p.m.n.; **SPD** m.p.m.n.; **J** p.n.; **AB**
March 4	(a) ADAMS (H.J.)† (b) – (e) ANON (f) [BLENKARN (S.A.)] (g) ANON	(a) – (b) Ool (c) Birds (d) HH (e) Min Fos Sh (f) Br col (g) Lib	Stevens 433 lots 24pp.	**J** f.p.f.n.; **N**; **BMNHE**
March 13	PROCTOR (Major F.W.), Part II	Western Pal ool	Stevens 254 lots 24pp.	**J**; **BMNHZ**; **N** m.p.s.n.
September 30	(a) – (g) ANON	(a) Min (b) Br ool For ool (c) Lib (d) Mam (e) Exotic lep (f) Br lep For lep Br col For col (g) Br lep Nh	Stevens 393 lots 28pp.	**J** s.p.s.n.; **BMNHE**; **N**
November 25	(a) SIMMONS (C.H.)† (b) NICOLSON (W.E.) (c) STANDEN (R.S.) (d) ANON (e) HARWOOD (W.H.), Colchester (f) ANON	(a) Br lep (b) European butts (c) European butts (d) Lib (e) Lep (*Abraxas grossulariata*) (f) Exotic lep	Stevens 436 lots 26pp.	**LC** m.p.m.n.; **J** m.p.m.n.; **CH** **BMNHE**
1914 January 13	(a) LOAT (W. Leonard S.) (b) – (k) ANON (l) TENNANT	(a) – (b) Birds (c) Birds Ool (d) – (g) Ool (h) Birds Fish (i) Ool (j) Sh Min (k) HH (l) Min Fos	Stevens 379 lots 19pp.	**J** f.p.f.n.; **BMNHZ**; **N**; **JAG**
February 10	(a) RAYNOR (Rev. G.H.) (b) JUPP (B.E.), Haslemere† (c) ANON	(a) Br lep (b) Br moths (c) Exotic lep	Stevens 396 lots 18pp.	**CH** m.p.m.n.; **J** m.p.m.n.; **LC** m.p.m.n.; **BMNHE**
February 24	(a) RATTRAY (Col. R.H.) (b) ANON (c) SLATER (Rev. H.A.)† (d) OWSTON (Alan), Yokohama (e) – (g) ANON	(a) – (e) Ool (f) Lib (g) Ool	Stevens 396 lots 27pp.	**BMNHZ**; **J**; **N** m.p.m.n.
February 25	HAZELDINE (Kendall), The Orchard, Woldingham, Surrey	Min Birds Ool	Sotheby 106 lots 9pp.	**BMNHM** p.m.n.; **BML** p.n.; **S** p.n. copy
April 7	(a) ANON (b) [?JEFFREYS (C.)] (c) ANON (d) RATTRAY (Col. R.H.) (e) MATHEW (G.F.) (f) MAITLAND (F.Lewis) (g) ANON (h) WHEELER (Edwin)†	(a) Exotic lep Col (b) Ool (c) Birds (d) Ool (e) Australian birds (f) Icelandic birds (g) Sh Min (h) Herb	Stevens 488 lots 22pp.	**BMNHZ**; **BMNHE**; **J** f.p.f.n.; **N** p.f.n.

1914–1916	SOURCE	CONTENTS	AUCTIONEER & SALE CAT.	REFERENCE
April 28-29	(a) MATHEW (G.F.) (b) ANON	(a) Br lep (b) Br lep Exotic lep	Stevens 673 lots 35pp.	J m.p.m.n.; CH m.p.m.n.; BMNHE
May 12	(a) – (c) ANON (d) LAYTON (Thomas)† (e) [WHEELER (Edwin)] †	(a) Exotic lep (b) Birds Mam (c) Sh (d) Fos Min (e) Herb	Stevens 419 lots 18pp.	J m.p.m.n.; CH BMNHE
June 16	WILKINSON (W.A.), Part I	Br ool	Stevens 273 lots 16pp.	J m.p.m.n.; N m.p.f.n.
June 30	(a) THRELFALL (J.H.)† (b) CAPPER (C.)	(a) Br microlep (b) Br lep	Stevens 414 lots 26pp.	J p.n.; LC m.p. m.n.; CH; BMNHE
July 14-15	(a) STUBBS (Rev. C.)† (b) – (e) ANON	(a) Br lep (b) Br lep Exotic lep (c) Birds Ool (d) Nh (e) Indian hh Indian mam	Stevens 620 lots 31pp.	J s.p.s.n.; BMNHE; N p.f.n.
October 20	(a) [ROSENBERG] (b) ANON (c) ANON (d) [SCHMASSMANN]	(a) Exotic lep (b) Birds Ool (c) HH Mam (d) Exotic lep	Stevens 330 + 14 bis lots 16pp.	J m.p.m.n.; CH m.p.; BMNHE
1915 March 11	(a) MUNN (P.W.) (b) ANON (c) SMITH (William H.)† (d) ANON	(a) – (d) Ool	Stevens 367 lots 20pp.	J; N m.p.f.n.
June 8	(a) [DEMANCHA (J.)] (b) – (c) ANON (d) WARNE (W.F.)[1] (e) – (g) ANON (h) SMITH (William H.)† (i) ANON	(a) Br lep (b) Br col For col (c) Exotic col (d) Br lep For lep (e) Indo-Australian butts (f) Birds (g) – (h) Ool (i) Sh	Stevens 534 lots 27pp.	J s.p.s.n.; BMNHE; N f.p f.n.
October 12	(a) ANON (b) [MILLS (H.O.)] (c) [WESTLAND (Wm.)†] (d) – (e) ANON	(a) Exotic lep (b) Br lep (c) Ool (d) Ool (e) HH	Stevens 370 lots 19pp.	J s.p.s.n. (incl. all Mills' lots); BMNHZ; BMNHE; CH; N
December 7	(a) ANON (b) [SCHMASSMANN] (c) ANON (d) [PRESTON (H.B.)]	(a) Br moths (b) Exotic lep (c) Nh (d) Sh (incl. some from J.T.MARSHALL & J.MABILLE colls.)	Stevens 327 lots 19pp.	NMW; CH; BMNHE; J
1916 April 4	(a) ANON (b) GRIST (Chas. J.) (c) – (d) ANON	(a) Br lep Exotic lep (b) Dup exotic lep (c) HH Mam (d) Nh	Stevens 397 lots 18pp.	J f.p.f.n.; CH; BMNHE
April 18	(a) WATERHOUSE (F.H.) & WATERHOUSE (E.A.)† (b) RAYNOR (G.H.) (c) MATHEW (G.F.) (d) – (e) ANON	(a) Br lep (b) Lep (*Abraxas grossulariata*) (c) Dup lep (d) Exotic lep (e) HH	Stevens 389 lots 19pp.	J m.p.m.n.; GS; CH; BMNHE

[1] Including 'the specimens figured in South's *Butterflies and Moths of the British Isles.*'

1916–1918	SOURCE	CONTENTS	AUCTIONEER & SALE CAT.	REFERENCE
June 20-21	(a) MANDERS (Col. Neville)† (b) CARBONELL (John)† (c) NOAKES (A.)† (d) [JANSON] (e) – (f) ANON (g) [DICKSEE]	(a) Br butts For butts (b) Lep Ool (c) Exotic lep (d) Books from Miss ORMEROD lib (e) Sh (f) HH (g) Exotic lep	Stevens 574 lots 31pp.	J m.p.m.n.; CH; BMNHE; N
November 7	(a) [SAUZE (A.), Sydenham] (b) – (e) ANON (f) PADDOCK (G.F.)†	(a) Br lep (b) Exotic lep (c) HH (d) Sh (e) Lib (f) Birds Ool	Stevens 430 lots 22pp.	J m.p.m.n.; CH; BMNHE; N f.p.
December 5	(a) TAUTZ (Percy H.)† (b) – (c) ANON	(a) Br lep (b) App (c) Exotic lep	Stevens 273 lots 15pp.	J m.p.m.n.; CH; BMNHE
1917 March 13	PROCTOR (Major F.W.)	Ool	Stevens 395 lots 31pp.	J f.p.f.n.; N p.n.
March 27	(a) KEMPT (W.) (b) HODGSON (Rev. A.E.)† (c) [HARWOOD (W.H.) & SON] (d) [DICKSEE] (e) ANON (f) [SCHMASSMANN]	(a) Br lep (b) Br lep (c) Lep (*Abraxas grossulariata*) (d) Exotic lep (e) Nh (f) Exotic lep	Stevens 422 lots 23pp.	J m.p.m.n.; CH; BMNHE
May 15	(a) RICHARDS (Percy) (b) – (c) ANON	(a) Br lep (b) Exotic lep (c) Ool	Stevens 424 lots 26pp.	J m.p.m.n.; CH f.p.; BMNHE; N f.p.
June 26	(a) GIBBS (A.E.)† (b) ANON	(a) American lep Pal lep (b) Lib	Stevens 434 lots 28pp.	J p.n.; CH f.p.; BMNHE
October 16	(a) BRIGGS (T.H.) (b) ANON	(a) Br lep (b) Exotic lep	Stevens 427 lots 28pp.	J m.p.m.n.; CH GS; BMNHE; SAR
November 13	(a) SCOLLICK (A.J.)† (b) [BUNBURY (Capt.)] (c) – (f) ANON	(a) Br lep (b) Br lep Exotic lep (c) Birds Ool (d) Lib (e) HH (f) Sh Min Fos	Stevens 328 + 50 bis lots 23pp.	J m.p.m.n.; CH s.p.s.n.; BMNHE; N; SAR
1918 February 26	(a) WOODROFFE (Rev. Duncan)† (b) – (e) ANON	(a) Fos Min Irr (b) Exotic col (c) Exotic lep Br lep Mam Sh Nh (d) HH (e) Mam	Stevens 257 lots 13pp.	J s.p.s.n.; CH; BMNHE
March 12	(a) RAYNOR (G.H.) (b) [MANGER] (c) WALL-ROW (T.) (d) ANON	(a) Lep (*Abraxas grossulariata*) (b) Br lep Exotic lep (c) – (d) Ool	Stevens 480 + 20 bis lots 28pp.	J m.p.m.n.; BMNHE; CH; N s.p.s.n.
May 14	(a) ALLEN (J.E.R.)† (b) [BUNBURY (Capt.)] (c) – (g) ANON	(a) Br lep (b) Exotic ins (c) Exotic lep (d) Sh (e) Sh Min Fos (f) Lib HH Birds Nh (g) Birds Nh	Stevens 380 + 6 bis lots 22pp.	J m.p.m.n.; CH s.p.; BMNHE
July 16	(a) JONES (Professor F. Wood) (b) ANON (c) [JENNER]	(a) Ool (b) HH Min Fos (c) Lib	Stevens 314 lots 24pp.	J; N m.p.s.n.
October 29	(a) GRANT (General Seafield)† (b) CARDEW (Lt.Col. P.A.) (c) – (f) ANON	(a) – (b) Br lep (c) Lib (d) Sh Fos (e) Exotic ins (f) HH Mam Nh	Stevens 322 + 1 bis lots 18pp.	J m.p.m.n.; CH; LC, BMNHE

1918–1919	SOURCE	CONTENTS	AUCTIONEER & SALE CAT.	REFERENCE
December 10	BRIGHT (Percy M.) [It seems this sale never took place, since both the Janson and BMNHE cats. are marked 'cancelled']	Br lep (Noctuae and *Abraxas grossulariata*)	Stevens 479 + 1 bis lots 27pp.	J; CH; BMNHE
1919 N.d.	PANKHURST [Sale unconfirmed]	Br lep	Stevens	Horn & Kahle (1936:201)
February 11	OLIVER (G.B.)	Br lep	Stevens 394 lots 19pp.	J m.p.m.n.; BMNHE
February 25	(a) CORNFORD (Rev. E. Bruce)† (b) – (c) ANON (d) [MORLEY (Wm. A.)]	(a) Br lep Exotic lep (b) Exotic lep (c) App (d) Lib	Stevens 441 lots 18pp.	J m.p.m.n.; BMNHE
March 11	(a) JOY† (b) ADAMS (F.C.) (c) – (f) ANON	(a) Br lep (b) Br col (c) Exotic lep (d) Ool (e) HH (f) Sh	Stevens 537 lots 26pp.	J s.p.s.n.; BMNHE; N
May 20	(a) COOKE (O.F.E.) (b) NORGATE (F.) (c) HOLMES (Mrs. C.)	(a) Br lep Exotic lep (b) Br macrolep (c) Br lep	Stevens 427 lots 23pp.	J p.n.; CH; BMNHE
June 4	(a) ANON (b) [HUISH] (c) ANON (d) – (i) ANON	(a) Br lep (b) Exotic ins (c) Lep (*Abraxas grossulariata*) bred by B.S.HARWOOD (d) Chinese butts (e) Lib (f) HH (g) Birds Ool (h) Sh Min Fos (i) Exotic lep	Stevens 493 lots 25pp.	J s.p.s.n.; CH; BMNHE
August 11–12	BALSTON (W.E.), Barvin, Potters Bar†	Lep Sh Fos Mam Nh, 4	Fairbrother, *house sale* 782 lots 40pp.	BMNHL
September 23	(a) BARRAUD (P.J.) (b) ANON (c) ROBERTSON (C.M.)† (d) – (e) ANON (f) PITMAN† (g) – (i) ANON	(a) Br macrolep (b) European butts (c) Br macrolep (d) Br lep (e) Br lep (f) Br macrolep (g) Birds (incl. Great Auk) Ool (h) Min Sh (i) HH	Stevens 476 lots 31pp.	J m.p.m.n.; CH; BMNHE; H; N
October 13–17, 20–21	SWANSEA (Lord), Singleton Abbey, Swansea	13 x: Birds Ins HH Min Cor Nh, 32	Knight, Frank & Rutley, *house sale* 1890 + 11 bis lots 116pp.	BMNHM p; VA p.n.
October 21	WEBB (Sydney)†, Part I	Br lep	Stevens 366 lots 20pp.	J p.n.; CH m.p. m.n.; LC; BMNHE; H; MAR p.n.
November 11	(a) MITFORD (R.S.) (b) – (e) ANON	(a) – (b) Br lep (c) – (d) Exotic lep (e) HH Mam	Stevens 395 lots 17pp.	J m.p.m.n.; CH f.p.; LC; BMNHE
December 9	WEBB (Sydney)†, Part II	Br lep	Stevens 422 lots 20pp.	J p.n.; CH m.p. m.n.; LC; BMNHE; MAR m.p.m.n.

1920–1921	SOURCE	CONTENTS	AUCTIONEER & SALE CAT.	REFERENCE
1920 January 20	(a) HALL (T.W.) (b) – (f) ANON (g) WEBB (Sydney)†	(a) Br lep (b) Br lep (c) Br lep Exotic lep (d) Col (e) Min (f) Mam HH Nh (g) Herb Sh Min Ool Zoophytes Br algae Br musci Br hepatics	Stevens 479 lots 22pp.	J m.p.m.n.; LC; BMNHE
February 10	WEBB (Sydney),† Part III	Br lep	Stevens 350 + 1 bis lots 19pp.	CH p.n.; J p.n.; BMNHE; LC; SAR
March 9	WEBB (Sydney),† Part IV	Br lep Lib	Stevens 334 + 2 bis lots 18pp.	CH p.n.; J p.n.; LC; BMNHE; H; SAR
March 23	(a) GARDNER (J.), Oxford St., London†, Part I (b) BILLUPS (T.R.)† (c) – (e) ANON (f) OGILVIE-GRANT (W.R.)	(a) Br lep Col Ool Nh (b) Br ins (c) Lib (d) Sh Cab (e) Birds Ool (f) Mam HH	Stevens 415 + 9 bis lots 22pp.	J m.p.m.n.; CH; BMNHE; H; SAR
April 20	(a) BOWEN-ROBERTSON (Major R.), Chandler's Ford, Hants† (b) GARDNER (James)†, Part II (c) ANON	(a) Br macrolep Br microlep (b) Ool Birds (c) Lib Nh	Stevens 460 lots 23pp.	J f.p.f.n.; CH; LC; BMNHE; H; SAR
May 18	(a) BRADY (L.S.) (b) ANON (c) GARDNER (James)†, Part III (d) ANON	(a) Br lep (b) Br macrolep Br microlep (c) Ool Birds Nh (d) Mam Birds Ool	Stevens 391 + 5 bis lots 19pp.	J s.p.s.n.; CH; BMNHE; SAR
June 22	(a) BATTLEY (A.U.)† (b) ANON (c) GORHAM (Rev. H.S.) (d) ANON (e) GARDNER (James)†, Part IV (final)	(a) Br macrolep Br microlep (b) Lep (c) Lib (d) Ool HH (e) Birds Mam HH Nh	Stevens 461 + 1 bis lots 21pp.	J f.p.f.n.; CH; LC; BMNHE; N; SAR
October 12	(a) – (b) ANON (c) [BRYANT] (d) GORHAM (Rev. H.S.)†	(a) – (b) Exotic lep (c) Br col (d) Exotic col	Stevens 561 + 5 bis lots 22pp.	J m.p.m.n.; CH m.p.; BMNHE; SAR
October 13	(a) FIELDEN (Colonel) (b) NOBLE (H.) (c) COTTON (T.A.) (d) ANON	(a) – (d) Lib	Stevens 308 + 1 bis lots 21pp.	CH; BMNHE; N s.p.
October 26	(a) ANON (b) LAWSON, Perth† (c) GWATKIN-WILLIAMS (Capt.) R.N. (d) – (e) ANON (f) KENNARD (M.T.)† (g) ANON	(a) – (d) Br lep (e) Lib (f) Mam HH (g) Birds Nh	Stevens 379 lots 19pp.	CH; BMNHE; J; LC; SAR
November 23	(a) BETHELL (Capt. the Hon. R.) (b) TURNER (E.)	(a) Birds Ool (b) Ool	Stevens 437 lots 30pp.	J; N s.p.f.n.; SAR
December 9–10, 13–15	NICKELS (Walter L.), Chenotrie Noctorum, Birkenhead†, Part II	9 xii: Min Fos Mam Cab, 4 14 xii: Fish Sh Nh, 3	Corkhill & Job, *house sale* 1000 (incl. 7 blank) lots 60pp. (41-100)	BMETH
1921 January 18	(a) MARSHALL (William), F.E.S.† (b) – (c) ANON (d) GRIST (C.J.) (e) – (h) ANON	(a) Br macrolep Br microlep (b) – (c) Br lep (d) – (e) Exotic lep (f) Min (g) Birds Ool Nh (h) HH Mam Lib	Stevens 509 + 10 bis lots 25pp.	J m.p.m.n.; CH m.p.m.n.; LC; BMNHE; SAR

1921–1922	SOURCE	CONTENTS	AUCTIONEER & SALE CAT.	REFERENCE
March 1 - 4	GODMAN (F. du Cane)	3-4 iii: Lib, 510	Sotheby 1017 lots 110pp.	J p.n.
March 15	(a) BOWER (B.A.), Chislehurst†, Part I (b) – (d) ANON	(a) Br lep (b) – (c) Exotic lep (d) Birds HH Nh Sh Min Ool Lib	Stevens 481 + 4 bis lots 23pp.	J m.p.m.n.; CH; LC; BMNHE; H SAR
May 10	(a) – (e) ANON (f) ROSEVEAR (J. Burman), Memb. Conchological Soc.† (g) ANON (h) BONHOTE (J.L.) (i) PIDSLEY (j) – (k) ANON	(a) Br lep (b) Br ins (c) Ins (d) Ins Lib (e) Sh Min (f) Sh Fos Min (g) Ool (h) Dup ool (i) Ool Nh (j) Ool (k) HH Mam Nh	Stevens 453 + 31 bis lots 24pp.	J m.p.m.n.; CH; LC; BMNHE; SAR
October 18	(a) HUDD (A.E.) (b) ANON (c) [HORSLEY (Cannon J.W.)] (d) DUCKWORTH (H.)† (e) – (f) ANON	(a) Br lep Exotic col (b) Exotic lep Br lep (c) Sh (d) Fos Min Nh (e) Ool Nh (f) HH Mam Nh	Stevens 528 + 34 bis lots 24pp.	CH s.p.s.n.; J f.p.f.n.; BMNHE; LC; SAR
1922 January 10-11	(a) – (d) ANON (e) FARN (A.B.)† (f) MILBURN (Major W.H.), Weybridge, Surrey (g) ANON (h) GREEN (J.F.) (i) FARN (A.B.)†	(a) Br lep (b) Exotic lep (c) Exotic col (d) Exotic lep (e) Lib (f) – (h) Ool (i) Ool Birds	Stevens 835 + 27 bis lots 44pp.	CH m.p.m.n.; J s.p.s.n.; BMNHE H; N m.p.m.n.; SAR
February 14	FARN (A.B.)†, Part I	Br lep	Stevens 400 + 1 bis lots 21pp.	CH p.n.; J m.p. m.n.; BMNHE; LC; SAR
March 14	FARN (A.B.)†, Part II	Br lep	Stevens 434 + 1 bis lots 21pp.	CH p.n.; J m.p. m.n.; BMNHE; LC; H; SAR; MAR m.p.m.n.
March 15	CHAPMAN (Dr. T.A.)†	Lib (mainly entomological)	Stevens 386 + 6 bis lots 25pp.	J p.n., CH s.p. s.n.; LC; BMNHE; SAR
April 4	(a) FARN (A.B.)†, Part III (final) (b) EDDRUP (Rev. T.B.)	(a) – (b) Br macrolep Br microlep	Stevens 353 + 1 bis lots 20pp.	J s.p.s.n.; CH m.p.m.n.; LC; BMNHE; H; SAR
May 16	(a) COOPER (S.)† (b) RUSSELL (J.W.) (c) TODD (R.G.) (d) – (h) ANON	(a) – (c) Br lep (d) Exotic lep (e) Br lep Exotic lep (f) Herb (g) Birds Mam (h) Lib	Stevens 477 + 28 bis lots 24pp.	CH m.p.m.n.; J m.p.m.n.; LC; BMNHE; H; SAR
June 13	(a) JAGER (J.)† (b) GIBB (Lachlan)† (c) – (f) ANON	(a) – (b) Br lep (c) Lep (d) Exotic ins (e) Exotic lep (f) Birds Herb Ool Sh Cor Fos Nh	Stevens 381 + 66 bis lots 21pp.	J m.p.m.n.; CH; LC; BMNHE; H; SAR
September 19	(a) PERKINS (V.R.)† (b) ANON (c) WATKINS & TULLETT, entomological dealers	(a) Br ins (Lep Col Hym Hem. Od Trichoptera) Exotic lep (b) – (c) Exotic lep	Stevens 496 + 27 bis lots 22pp.	J p.n.; CH s.p.; LC; BMNHE
November 21	(a) DAY (Rev. A.) (b) ANON (c) SWINHOE (Ernest)† (d) – (f) ANON	(a) Br macrolep Br microlep (b) Lep (*Abraxas grossulariata*) (c) Exotic butts (his stock of) (d) Lib (e) Br lep (f) Br col	Stevens 670 + 20 bis lots 32pp.	CH f.p.f.n.; J f.p.; LC; BMNHE; SAR

1923–1924	SOURCE	CONTENTS	AUCTIONEER & SALE CAT.	REFERENCE
1923 January 30	HORNE (Arthur), Aberdeen†, Part I	Br butts	Stevens 538 lots 23pp.	**J** p.n.; **CH** m.p. m.n.; **LC**; **BMNHE**; H; SAR; MAR p.n.
February 20	HORNE (Arthur), Aberdeen†, Part II	Br macrolep	Stevens 398 lots 18pp.	**J** p.n.; **CH** m.p. m.n.; **BMNHE**; **LC**; H; SAR
April 10	(a) HORNE (Arthur), Aberdeen†, Part III (final) (b) ANON	(a) Br macrolep (b) Br lep Lib	Stevens 436 lots 18pp.	**CH** m.p.m.n.; **J** m.p.m.n.; **LC**; **BMNHE**; AB f.p.; SAR
May 29	(a) NEWNHAM (F.B.)† (b) – (f) ANON (g) FARWELL (C.G.) (h) ANON	(a) Pal lep (b) – (d) Br lep (e) Br butts (f) Br lep Exotic lep (g) Exotic lep (h) Sh Mam	Stevens 601 + 11 bis lots 27pp.	**J** m.p.m.n.; **CH** s.p.; **LC**; **BMNHE**; AB f.p.; SAR
August 29	BULOW (Carl) [Sale unconfirmed]	Sh	Stevens	Tomlin (1941: 166)
October 9	ANON	Br lep Exotic lep Col Hem App Cab	Stevens 438 lots 16pp.	**J** f.p.f.n.; **CH** f.p.; **LC**; **GS**; **BMNHE**; SAR
November 20	(a) TERRY (Major H.A.)† (b) ANON (c) ANON (d) CRUTTWELL (Canon C.T.) (e) ANON (f) [ESSON] (g) ANON	(a) Ool (b) Birds (from F.BOND coll.) Ool (c) Exotic lep Exotic col App[1] (d) Br macrolep Br microlep (e) Br lep Br ool (f) Br lep (g) Br lep Nh	Stevens 616 lots 32pp.	**J** m.p.m.n.; **CH** f.p.; **LC**; **BMNHE**; **N** p.m. n. (ool only); SAR
December 11	(a) PREST (E.E.B.) (b) – (e) ANON	(a) Br macrolep (b) Br butts (c) Br macrolep (d) Br lep For lep (e) Exotic lep	Stevens 644 + 2 bis lots 25pp.	**J** m.p.m.n.; **CH** s.p.s.n.; **LC**; **BMNHE**; SAR
1924 January 22	ANON	Exotic butts Exotic ins App Cab	Stevens 429 lots 16pp.	**CH** m.p.f.n.; **J** f.p.f.n.; **GS**; **BMNHE**; SAR
February 26	(a) ELWES (H.J.), F.R.S.† (b) FRANCIS (Dr. W.)† (c) CLARK (S.V.), Angmering† (d) – (g) ANON	(a) – (c) Ool (d) Birds (e) Br lep Exotic lep (f) Exotic lep (g) Sh	Stevens 524 + 11 bis lots 27pp.	**J** s.p.s.n.; **CH** s.p.; **N** p.m.n.; **BMNHE**; AB f.p; SAR
March 25	(a) HEMMING (A.F.) (b) THORNHILL (E.H.) (c) – (g) ANON	(a) Br butts (b) Br macrolep (c) Exotic butts (d) Lep (e) Exotic butts (f) Sh Herb Nh (g) Lib	Stevens 445 lots 23pp.	**J** m.p.m.n.; **CH** s.p.f.n.; AB f.p.; **BMNHE**; SAR; MAR m.p.m.n.
April 8	ANON	Exotic lep App	Stevens 433 lots 16pp.	**GS** m.p.; **J**; **BMNHE**; **CH** copy, SAR

[1] Included lot 447: 'A large specimen cabinet, bearing an inscription stating that it was made in the year 1756 by Linneus for his own use.'

1924–1925	SOURCE	CONTENTS	AUCTIONEER & SALE CAT.	REFERENCE
June 3,5	(a) – (g) ANON (b) ELWES (H.J.)†	(a) Exotic butts (b) App Ins (c) Br lep (d) App (e) Br lep (f) Br lep (g) Exotic ins (h) Lib Birds	Stevens 1087 + 13 bis lots 46pp.	J s.p.s.n.; CH s.p.; LC; GS; BMNHE; SAR
July 1	(a) ANON (b) [SCOTT (J.W.)] (c) – (g) ANON	(a) Exotic butts Ins (b) Br macrolep (c) Ool (d) Birds (e) Birds Rept (f) Sh (g) Lib	Stevens 566 + 3 bis lots 26pp.	CH s.p.; GS; LC J; BMNHE; N; SAR
July 14	CARNEGIE (David John, 10th Earl of Northesk)†	Bot (objects from Bronze Age Swiss Lake Dwellings), ?	Christie 709 lots 66pp.	BMETH p.n.
October 7	(a) – (b) ANON	(a) Lep (b) App	Stevens 456 lots 17pp.	CH f.p.; J; GS; BMNHE; SAR
October 28	(a) – (f) ANON	(a) App (b) Exotic lep (c) Ool (d) Birds App Lep Ins (e) Lib (f) Br butts	Stevens 599 + 2 bis lots 26pp.	J f.p.f.n.; CH; GS; BMNHE; SAR
December 2	(a) DALGLISH (A.A.), Glasgow† (b) LLOYD (A.), F.E.S., F.C.S.† (c) STYAN (T.G.) (d) DOWNES (Rev. A.M.) (e) THORNEWILL (Rev. C.F.) (f) – (i) ANON	(a) Br ins (incl. Lep Col Hem Hym Neur Dipt) (b) – (c) Br macrolep Br microlep (d) Br lep (e) Br lep (f) Br lep For lep (g) App (h) Ool Cab (i) Lib	Stevens 490 + 3 bis lots 29pp.	J m.p.m.n.; CH s.p.f.n.; LC; GS; BMNHE; N
1925 February 10	(a) JONES (A.H.)† (b) – (d) ANON	(a) Br macrolep Br microlep (b) Lib (c) Br lep (d) Lep Nh	Stevens 360 + 2 bis lots 17pp.	CH m.p.f.n.; J s.p.s.n.; GS; LC; BMNHE; SAR
March 17	(a) GRIFFITHS (G.C.), F.E.S., Bristol† (b) ANON (c) JONES (A.H.) (d) RAYNOR (Rev. Gilbert (e) – (g) ANON	(a) Br lep (b) Exotic lep (c) For butts Exotic lep (d) Lep (*Abraxas grossulariata*) (e) – (f) Exotic lep (g) Lib	Stevens 615 lots 31pp.	J m.p.m.n.; CH m.p.f.n.; LC; GS; BMNHE; SAR
June 30	(a) MAY (Hubert)† (b) [BOOTH (H.T.)] (c) ANON (d) CANSDALE (F.E.)† (e) FAWCETT (Colonel)† (f) – (i) ANON	(a) – (b) Ool (c) Birds Ool Mam Nh (d) Br macrolep Br microlep Br col (e) Exotic lep (f) Br lep & larvae (g) Br lep (h) Lib (i) HH	Stevens 628 + 6 bis lots 34pp.	J m.p.m.n.; CH m.p.m.n.; GS; LC; BMNHE; N m.p.m.n. (ool only); SAR
September 29	(a) ABBOTT (S.) (b) ANON (c) SMALLEY (F.W.) (d) ANON	(a) Br macrolep (b) Br lep Exotic lep (c) Lib (d) Exotic lep	Stevens 533 + 23 bis lots 29pp.	J m.p.m.n.; CH m.p.f.n.; BMNHE; LC; GS; SAR; MAR s.p.
October 20	CREWE (Sir Vauncey Harper)†	Br lep (Part I)	Stevens 412 lots 19pp.	J p.n.; CH m.p. m.n.; LC; GS; BMNHE; N; SAR; MAR p.n. (of butts only)
November 10	CREWE (Sir Vauncey Harper)†	Br birds Mam Fish	Stevens 374 lots 23pp.	J m.p.m.n.; LC; BMNHE; N; SAR
November 24	(a) CREWE (Sir Vauncey Harper)† (b) ANON	(a) Br lep incl. Microlep (Part II & final) (b) Lep Orth Neur	Stevens 377 lots 20pp.	CH p.n. add; J m.p.m.n.; LC; GS; BMNHE; N; SAR; MAR p.n.

1925–1927	SOURCE	CONTENTS	AUCTIONEER & SALE CAT.	REFERENCE
December 15	CREWE (Sir Vauncey Harper)†	Br ool	Stevens 352 + 1 bis lots 23 + 1pp.	**J**; **LC**; **BMNHE**; **N** p.n.; **SAR**
1926 February 9	ANON	Min Fos, 28	Stevens 464 lots 20pp.	**J**
February 23	CREWE (Sir Vauncey Harper)†	Birds Ool Mam	Stevens 422 lots 21pp.	**J** m.p.m.n.; **LC**; **BMNHE**; **N**; **SAR**
March 2	(a) BROWN (Major E.W.), Cambridge (b) – (c) ANON (d) BUCKLEY (Dr. G.Granville), M.D., F.S.A., F.E.S. (e) – (i) ANON	(a) Br macrolep Br microlep (b) Br lep Cab Nh (c) Lib (d) Exotic lep (e) App Lep (f) Exotic lep Exotic col Herb Birds Nh (g) Exotic lep (h) HH (i) Exotic lep (incl. Microlep)	Stevens 581 + 1 bis lots 28pp.	**CH** f.p.f.n.; **LC**; **BMNHE**; **SAR**
March 9	(a) CROWFOOT (Dr. W.M.)† (b) ANON	(a) Br ool (b) Lib	Stevens 324 lots 24pp.	**J** f.p.f.n.; **N** m.p. m.n.; **BMNHE**; **SAR**
April 13	(a) CREWE (Sir Vauncey Harper)† (b) – (e) ANON (f) CHAMP (H.)† (g) WILLIAMS (J.M.)†	(a) Ool Birds Mam (b) Birds Nh (c) Lib (d) HH (e) Exotic lep (f) Sh (g) Sh	Stevens 530 lots 27pp.	**J** s.p.s.n.; **CH** s.p.; **BMNHE**; **N** m.p.m.n. (of ool only), **AB** f.p.; **SAR**
June 22	(a) FENN (Charles), F.E.S. (b) – (e) ANON	(a) Br macrolep Br microlep (b) Br ins (c) Exotic ins Br ins (d) Br ins Exotic ins Nh (e) Lib	Stevens 513 + 17 bis lots 26pp.	**CH** m.p.s.n.; **GS**; **LC**; **BMNHE**; **SAR**
November 23	(a) BUTLER (W.E.), F.E.S.† (b) ANON (c) St. BARTHOLO-MEW'S HOSPITAL (d) – (f) ANON (g) [LAWSON (P.)] (h) ANON	(a) Br macrolep (b) Col (c) Br lep Ins (Haworth types) (d) Br lep Cab App (e) Exotic butts (f) Ool Birds Sh Nh (g) Sh Min Fos (h) Lib Mam HH Nh	Stevens 321 + 1 bis lots 16pp.	**J** m.p.m.n.; **CH** m.p.m.n.; **GS**; **LC**; **BMNHE**; **N**; **AB** f.p.; **SAR**
1927 February 8	(a) DOEG (T.E.)† (b) FORSTER (W.)† (c) – (e) ANON	(a) Exotic lep Ool Lib (b) Birds Ool (c) Ool (mostly Irish) (d) Br lep Exotic lep (e) Lib	Stevens 554 + 15 bis lots 30pp.	**BMNHE**; **N** m.p.n. (of ool only)
May 31– June 1	(a) HIDEN (F.C.) (b) – (d) ANON (e) JACKSON (T.W.)† (f) NEVINSON (E.B.) (g) ANON (h) LINDEMANN (O.) (i) – (j) ANON	(a) – (b) Ool (c) Birds (d) Lib (e) Br macrolep Br microlep (f) Br macrolep Br microlep (g) Lep (h) Exotic lep (i) Exotic ins (j) Exotic butts	Stevens 844 lots 47pp.	**CH** s.p.s.n.; **GS**; **BMNHE**; **N** m.p. s.n. (ool only); **SAR**
October 11	(a) – (c) ANON (d) RAYNOR (G.H.) (e) ANON (f) MOORE (Major F.C.) (g) – (i) ANON	(a) Exotic butts (b) Exotic butts (c) Br lep Br col (d) Br lep (*Abraxas grossulariata*) (e) Birds Ool (incl. *Aepyornis maximus*) (f) – (i) Ool	Stevens 446 lots 26pp.	**J** m.p.m.n.; **CH** m.p.m.n.; **LC**; **GS**; **BMNHE**; **N** m.p.m.n. (ool only); **SAR**
November 8	(a) CLUTTERBUCK (C.G.) (b) – (f) ANON (g) [GYNGELL (W.)] (h) ANON	(a) Br lep (b) Br lep (c) For lep (d) Ool (e) Min (f) Lib (g) Sh (h) Fos Nh	Stevens 397 + 10 bis lots 22pp.	**LC** m.p.m.n.; **J** s.p.s.n.; **GS**; **BMNHE**; **CH** copy; **AB** f.p.; **SAR**

1928–1929	SOURCE	CONTENTS	AUCTIONEER & SALE CAT.	REFERENCE
1928 March 13	(a) VEREL (J.B.) Norwich† (b) KING (Dr. T.W.), Dorking† (c) MARSH (H.)† (d) ENTOMOLOGICAL CLUB (e) ANON	(a) Ool Lib (b) Br lep (c) Ins (d) Ins Cab (e) Mam Sh Nh	Stevens 482 lots 30pp.	**J** s.p.s.n.; **LC**; **BMNHE**; **N** m. m.n. (ool only) SAR
May 8	(a) REID (P.C.)† (b) SKINNER (Percy F.) (c) PORRITT (G.T.) (d) ANON (e) HARTING (J.E.) (f) [STEVENSON (Henry)]	(a) Br lep Lib (b) Br lep Br col (c) Lib (d) Exotic lep (e) – (f) Ool	Stevens 477 + 24 bis lots 32pp.	**BMNHE**; **GS** f.p.; **CH** copy; **N**; SAR
July 10	ANON	Mam Narwhal tusks Elephant tusks, 12	Stevens 360 lots 20pp.	**BMETH**
November 13	(a) CROSS (F.B.) (b) ST. JOHN (W.) (c) ANON (d) HOLDAWAY (A.E.) (e) JONES (C.W.B. Cuthbert) (f) – (g) ANON (h) COXON (H.)† (i) – (j) ANON (k) [RADLEY (P.)] (l) ANON	(a) – (c) Br lep (d) Br macrolep Br microlep (e) Br butts (f) Br lep Sh (g) – (h) Ool (i) Birds (j) Birds Ool Exotic lep (k) Sh (l) Lib	Stevens 562 lots 34pp.	**CH** s.p.s.n.; **LC** s.p.s.n.; **GS**; **BMNHE**; **N** s.p. s.n. (ool only); AB f.p.
1929 January 30– February 1	PEEK (Sir Wilfred), Rousden, Devon†	Sh Ool Fos Min Nh, 41	Glendining 730 lots 42pp.	**BMNHG**; **BMETH** m.p.
March 12	(a) RIDLEY (P.W.) (b) ANON (c) VICKERS (J.)† (d) ANON (e) SLOCOMBE (Shirley) (f) – (l) ANON (m) [DAMON (R.F.), Weymouth], et al (n) ANON	Br macrolep Br microlep (b) Pal butts (c) Exotic butts (d) Br ool (e) Ool Nests (f) Ool (g) Birds Nh (h) Exotic lep (i) Exotic lep (j) Birds (k) Lep (l) Exotic lep (m) Sh Fos (n) Sh Cor Min Fos	Stevens 543 +10 bis lots 30pp.	**J** m.p.m.n.; **CH** **GS**; **LC**; **BMNHE**; **N**
March 19-20	ANON	23 vii: Nh (few lots only) Irr	Stevens 638 lots 34pp.	**BMETH**
May 7	DALGLEISH (J.J.)†	Ool	Stevens 489 lots 33pp.	**BMNHE**; **LC**; **N**
June 11	NICHOLS (J.B.)†	Birds (incl. Great Auk)	Stevens 487 lots 36pp.	**J**; **BMNHE**; **N** m.p.m.n.; SAR
July 3	(a) DACIE (J.C.)† (b) TAYLOR (Thomas Lombe), Starston, Norfolk (c) ANON (d) DALGLEISH (J.J.)†	(a) Sh (b) Sh (final part) (c) Ins (d) Ool	Stevens 408 + 1 bis lots 27pp.	**NMW** m.p.m.n.; **SPD**; **BMNHE**; AB s.p.
October 9	(a) BAXTER (J.), Dundee† (b) LUDLAM (G.S.) (c) DALGLEISH (J.J.) (d) – (f) ANON	(a) – (d) Ool (e) Exotic lep (f) Lib	Stevens 557 lots 35pp.	**J**; **BMNHE**; **N** m.p.f.n.

1929–1931	SOURCE	CONTENTS	AUCTIONEER & SALE CAT.	REFERENCE
November 13	(a) NEVINSON (E.B.) (b) – (g) ANON (h) WRIGHT (Rev. H.J.)† (i) ANON (j) CORNELL (E.)† (k) ANON (l) PEACH (A.W.)[1] (m) RAYNOR (G.H.) (n) HARWOOD (B.S.)	(a) Br macrolep (b) Br col (c) Br lep (d) Br bats (e) Br sh Br starfish (f) Br lep Birds Ool (g) Br lep Lib (h) Br macrolep Br microlep (i) Ins Nh (j) Lep (k) Lep (l) Br lep (m) – (n) Lep (*Abraxas grossulariata*)	Stevens 590 + 23 bis lots 30pp.	CH m.p.m.n.; GS; LC; BMNHE
1930 February 12	JOHNSON (C.F.)†, Part I	Br lep	Stevens 371 lots 20pp.	CH p.n.; GS; LC; J; BMNHE
March 12	(a) JOHNSON (C.F.)†, Part II (final) (b) LONGHURST (A.M.) (c) ANON (d) MARSHALL (Eric)	(a) Br butts (b) Br macrolep (c) Exotic lep (d) For butts	Stevens 482 + 11 bis lots 28pp.	CH m.p.m.n.; GS; LC; J m.p. s.n.; BMNHE; MAR p.n.
March 28	(a) KING (J.), Holmwood, Surrey[2] (b) [FLEMING (J. McA.), Cambridge] (c) GYNGELL (W.), Scarborough	(a) – (b) Ool (c) Sh Cab	Stevens 513 + 11 bis lots 28pp.	BMNHE; N m.p. s.n. (of ool only) AB f.p.
May 7	(a) MELVILL (Dr. J. Cosmo), F.E.S.† (b) COOPER (B.) (c) ANON	(a) Br macrolep Br microlep (b) Br lep (c) Lep	Stevens 544 lots 28pp.	CH m.p.s.n.; GS; LC; BMNHE; MAR s.p.
June 18	(a) GARDNER (J.E.)† (b) SCORER (A.G.) (c) – (e) ANON	(a) Br macrolep Br microlep (b) Br lep (c) Exotic lep (d) Birds Ool (e) HH Cor	Stevens 559 + 3 bis lots 30pp.	LC m.p.s.n.; GS; BMNHE; CH copy
October 29	(a) REID (Captain Savile G.)† (b) TANCOCK (Canon O.W.)† (c) ANON	(a) Ool Br lep Lib (b) Ool Lib (c) Lep	Stevens 469 lots 28pp.	BMNHE; GS; CH; N p.m.n. (of ool only)
November 11-12	ROWLANDS (J.F.), late Vice Consul at Mollendo, Peru	Exotic lep Exotic col Nh, 10	Stevens 828 + 4 bis lots 39pp.	BMETH
November 12	RAYNOR (Rev. G.H.)†	Br lep For lep	Stevens 404 + 1 bis lots 19pp.	CH p.f.n.; LC; GS; BMNHE
December 3	(a) EDWARDS (Miss A.D.), East Grinstead† (b) EASTWOOD (J.E.), F.E.S. (c) BANKES (E.R.) (d) PICKLES (Frederick)	(a) Br macrolep (b) Br macrolep Br microlep Lib (c) Lib (d) Sh	Stevens 390 lots 23pp.	CH s.p.f.n.; LC; GS; BMNHE; AB f.p.
1931 May 12-13	(a) PARRIS (R. Stanway) (b) MEADEN (Louis) (c) ANON (d) [PARRITT (H.W.)] (e) ANON (f) SLOCOMBE (Shirley) (g) ANON (h) WOOD (C.R.) (i) HIDEN (F.C.)	(a) – (b) Br lep (c) Exotic lep (d) Japanese sh (e) Lib (f) Ool Nests (g) Ool (Great Auk) (h) – (i) Ool	Stevens 735 + 11 bis lots 44pp.	N s.p.s.n.; GS; BMNHE; CH copy; AB f.p.; LC s.p.f.n. inc., EG

[1] Also known by his *nom de plume* Arthur Valentine, under which name he had a further sale on 14.XI.1963.

[2] A ms. note by F.C.R.Jourdain in the Nichols cat. reads: 'All data in handwriting of P.F.Bunyard.'

1931–1934	SOURCE	CONTENTS	AUCTIONEER & SALE CAT.	REFERENCE
July 28-29	ANON	28 vii: Elephant tusks (lot 59), 1	Stevens 557 lots 33pp.	BMETH
August 11-12	ANON	11 viii: Mam HH, 11	Stevens 571 lots 28pp.	BMETH
November 11	(a) HOOD (Lord Viscount) (b) JACOBS (Major J.J.) (c) NEWNHAM (C.E.) (d) WALLROW (J.) (e) WINSTONE (Dr.)† (f) [COLTART (N.B.)]	(a) – (c) Br lep (d) Ool (Warblers) (e) Ool (f) Dup ool	Stevens 528 + 35 bis lots 30 + 1pp.	J s.p.s.n.; CH; LC; GS; BMNHE; N m. f.n.
1932 May 4	(a) DICKSEE (Arthur) (b) JOHNSON (E.E.)	(a) Exotic lep (b) Br lep	Stevens 540 lots 24pp.	CH m.p.; GS; LC; BMNHE; N s.p.
June 22	(a) SPERRING (C.W.) (b) HICK (Rev. J.M.)† (c) – (e) ANON (f) GYNGELL (W.), Scarborough	(a) Br lep (b) Br macrolep Br microlep Exotic butts (c) Lib (d) Exotic lep (e) Ool (f) Sh	Stevens 380 + 27 bis lots 22pp.	CH m.p.f.n.; GS; LC; J; BMNHE; N; AB f.p.
December 7	(a) RIDLEY (Mrs. P.W.) (b) ANON (c) PRICE (Lloyd)† (d) – (e) ANON	(a) Br lep (b) Pal lep Exotic lep (c) Birds Ool (d) Lib (e) Ool (*Aepyornis titan*)	Stevens 580 + 1 bis lots 30pp.	GS; BMNHE; N m.p.; CH copy; LC s.p. inc.
1933 May 3	(a) SNOWDEN (Canon A.H.) (b) MARCHANT (G.le)† (c) ANON (d) ANON (e) HARWOOD (B.S.) (f) CROWFOOT (Dr.)†	(a) – (b) Br lep (c) Br lep (*Arctia caja*) (d) Br lep (e) Dup Br lep (f) Ool Nests	Stevens 448 + 1 bis lots 26pp.	CH m.p.f.n.; GS; LC; BMNHE; N
October 25	(a) HAYWARD (J.H.)† (b) – (d) ANON (e) WILKINSON (W.A.)† (f) ANON (g) GEAR (A.T.)† (h) ANON	(a) Br lep (b) Br lep (*Arctia caja*) (c) Exotic lep (d) Br ool (e) Ool Nests (f) Mam Nh (g) Fish (coloured casts) (h) Lib	Stevens 415 + 12 bis lots 26pp.	CH m.p.; GS; LC; BMNHE; N s.p.
1934 March 27	(a) NURSE (Lt.Col. C.G.)† (b) CRAFT (J.G.) (c) PROUDFOOT (Rev. Samuel)† (d) QUIBELL (W.) (e) – (h) ANON	(a) – (c) Br lep (d) Br lep (*Abraxas grossulariata*) (e) Br lep (f) Exotic lep (g) Ool Nh (h) Lib	Stevens 504 + 21 bis lots 27pp.	CH m.p.; LC; GS; BMNHE; H
June 20	(a) LESTER† (b) ANON (c) ANON (d) BAILY (W.Shore)† (e) ANON	(a) Br lep (incl. larvae) (b) Sh (c) Exotic lep Nh (d) Ool (e) Ool	Stevens 342 + 12 bis lots 22pp.	CH; BMNHE; GS; N s.p.
November 14	ROWLEY (George Dawson), Brighton†	Birds Ool Lib	Stevens 382 lots 27pp.	BMNHZ; GS; N s.p.; WHB; EG
December 5	(a) MUTCH (J.P.)† (b) GREEN (J.)† (c) BAXTER (T.) (d) – (e) ANON	(a) Br macrolep Br microlep (b) – (c) Br lep (d) Br lep For lep (e) Exotic lep Sh Nh	Stevens 446 lots 26pp.	CH m.p.f.n.; GS; LC; BMNHE; BMNHZ; N; SAR

1935–1937	SOURCE	CONTENTS	AUCTIONEER & SALE CAT.	REFERENCE
1935 February 27	(a) ANDERSON (J.)†, Chichester (b) ANON (c) ANON	(a) Br lep Exotic lep (b) Ool (incl. *Aepyornis*) Birds (c) HH	Stevens 406 + 2 bis lots 22pp.	**CH** f.p.f.n.; **GS**; **LC**; **BMNHE**; **N**; **SAR**
March 6	CRABTREE (B.H.), Part I	Br macrolep	Stevens 467 + 2 bis lots 18pp.	**CH** p.m.n.; **GS**; **LC**; **BMNHE**
March 21	COLLIER (John)†	Min Cab Catalogue of coll. (by W.J.LEWIS ABBOTT, F.G.S., who formed the coll.), 1	Sotheby 154 lots 20pp.	**BML** p.n.; **S** p.n. copy
April 3	CRABTREE (B.H.), Part II	Br macrolep	Stevens 398 + 1 bis lots 16pp.	**CH** p.m.n.; **GS**; **LC**; **BMNHE**; **SAR**
October 22	(a) DAVIS-WARD (J.) (b) ANON (c) DEWEY, Eastbourne† (d) MITCHELL (A.T.)† (e) REUSS (C. Kennedy) (f) ANON (g) HADFIELD (James & Mrs.)	(a) Br lep Exotic lep Lib (b) Br lep (c) Br macrolep (d) Br macrolep (e) Lep (f) Birds Nh (g) Sh (from Lifu & adjacent isles)	Stevens 394 + 5 bis lots 23pp.	**CH** m.p.f.n.; **GS**; **LC**; **BMNHE**; **SAR**; **AB** f.p.; **MAR** f.p.f.n.
December 4	(a) GOODMAN (O.R.† & A.de B.) (b) – (e) ANON (f) RAYWARD (A.L.)† (g) ANON	(a) Br lep Pal lep Cab (b) Birds Ool (c) Br hym (d) Br dipt (e) European hym (f) Br macrolep Br microlep (g) Lib	Stevens 342 + 2 bis lots 20pp.	**CH** m.p.f.n.; **LC**; **GS**; **BMNHE**; **N**; **SAR**
December 16-17	ADKIN (Robert)†	17 xii: Entomological lib Irr	Sotheby 530 lots 69pp.	**LC** p.
1936 February 11	(a) NASH (Dr. Gifford), Bedford† (b) ANON	(a) – (b) Br lep	Stevens 300 + 5 bis lots 14pp.	**CH** p.m.n.; **GS**; **LC**; **BMNHE**; **MAR** m.p.m.n.
March 25	(a) WICKHAM (Rev. Prebendary) (b) [BROWN (E.W.)] (c) – (i) ANON	(a) Lep (b) Br microlep (c) Exotic lep (d) Br lep (e) Exotic lep (f) Exotic col (g) Exotic mam (h) Exotic birds (i) Min Fos Sh Nh	Stevens 382 lots 20pp.	**CH** p.f.n.; **LC**; **GS**; **BMNHE**; **N**; **SAR**
November 10	(a) WILLIAMS (C.H.)† (b) PEARMAN (Capt. A.) (c) ANON	(a) Br lep (b) Lep (c) Br lep	Stevens 419 + 2 bis lots 19pp.	**CH** p.m.n.; **GS**; **LC**; **BMNHE**; **SAR**
December 8	(a) MORRIS (J.B.) (b) NOBLE (Heatley)†	(a) Br lep (b) Ool Birds (young in down) Lib	Stevens 420 + 10 bis lots 20pp.	**CH** f.p.f.n.; **GS**; **LC**; **SAR**; **BMNHE**; **N** s.p.; **WHB**
December 9	PETHER (Wm. G.)	Br macrolep Lib	Stevens 516 lots 23pp.	**CH** m.p.; **GS**; **LC**; **BMNHE**; **SAR**; **MAR** p.n. (of butts only)
1937 February 16	(a) HARWOOD (B.S.) (b) ANON	(a) Br lep Ins Lib (b) Br lep Lib Exotic lep Ins	Stevens 480 lots 23pp.	**CH** m.p.m.n.; **LC** m.p.s.n.; **GS** s.p.; **BMNHE**; **SAR**; **MAR** m.p.m.n.

1937–1939	SOURCE	CONTENTS	AUCTIONEER & SALE CAT.	REFERENCE
March 16-17	MASSEY (Herbert), Parts I & II	Br macrolep	Stevens 1000 + 2 bis lots 36pp.	**CH** p.m.n.; **GS; LC; BMNHE;** SAR; MAR f.n.
April 6	(a) MASSEY (Herbert), Part III (final) (b) STIFF (Rev. Alfred)†	(a) Br macrolep (b) Br macrolep (moths only)	Stevens 292 + 5 bis lots 12pp.	**CH** p.m.n.; **GS; LC; BMNHE;** AB f.p.; SAR
April 20-21	(a) LOYD (L.R.W.) (b) – (d) ANON (e) McCLELLAND (H.), Birmingham	(a) Ool Cab (b) Lib (c) Ool Sh Ins (d) Br birds For birds Br ool For ool Ins (e) Sh	Stevens 714 lots 36pp.	**NMW** s.p.; **BMNHZ; BMNHE; GS** s.p. **LC; N** m.p.; SAR
1938 January 25	(a) – (e) ANON	(a) Exotic lep (b) Exotic ins (incl. Col Hem Od) (c) Exotic lep Ins (d) Br col Lep (e) Br lep	Stevens 505 + 6 bis lots 20pp.	**CH** m.p.; **GS; LC; BMNHE;** SAR
February 8	(a) BROCKLEHURST (W.S.) (b) – (c) ANON	(a) Br macrolep Br microlep (b) – (c) Br lep	Stevens 377 + 7 bis lots 16pp.	**CH** p.s.n.; **GS; LC; BMNHE;** SAR; MAR m.p m.n.
March 8	BRIGHT (Percy M.)	Br butts (part of coll.)	Stevens 284 lots 12pp.	**CH** p.n.; **GS; LC BMNHE;** SAR; MAR p.m.n.
May 11	(a) [CORDER (J.W.)] (b) – (h) ANON	(a) Br macrolep (b) Lep (larvae) Ins (c) Exotic lep (d) Ool (e) Ool Birds (f) Br ool (g) Sh (h) Ool	Stevens 431 + 5 bis lots 24pp.	**CH** s.p.f.n.; **GS; LC; BMNHE; N** s.p.; SAR
October 25-26	BOUCK (Baron)†	Br macrolep Br microlep (incl. life-histories, larvae, pupae, hymenopterous parasites)	Stevens 999 + 3 bis lots 38pp.	**CH** p.m.n.; **BMNHE; LC; GS; H;** SAR; MAR p.n.
November 15-16	HANBURY (F.J.)†	Br macrolep Br microlep	Stevens 814 + 5 bis lots 28 + 1pp.	**CH** s.p.s.n.; **GS; LC; BMNHE; H;** SAR; MAR p.n.
December 13	(a) BOUCK (Baron)† (b) McLACHLAN (R.)† (c) BALDOCK (G.R.)†	(a) – (b) Br lep (c) Br lep Exotic lep	Stevens 499 + 2 bis lots 23pp.	**CH** m.p.f.n.; **GS LC; BMNHE; H;** SAR
1939 February 7	BUNYARD (Percy F.) Part I	Ool	Stevens 392 lots 16pp.	**GS; N** m.p.m.n. **CH** copy
February 14	THOMPSON (J. Antony)	Br macrolep	Stevens 395 + 7 bis lots 14pp.	**CH** p.m.n.; **GS; LC; BMNHE; H;** SAR; MAR p.n.
March 7	BUNYARD (Percy F.), Part II	Ool	Stevens 333 lots 16pp.	**GS; N** p.m.n.; **CH** copy

1939–1943	SOURCE	CONTENTS	AUCTIONEER & SALE CAT.	REFERENCE
March 14	(a) WALKER (S.), York† (b) PICKETT (C.P.)† (c) GREER (Thomas)	(a) Br macrolep (b) Br lep Exotic lep Ins (c) Butts	Stevens 437 + 2 bis lots 19pp.	**CH** p.m.n.; **GS**; **LC**; **BMNHE**; H; SAR; MAR m.p.m.n.
April 25-26	(a) GILLES (W.S.)† (b) ANON (c) TUNALEY (H.)† (d) FOX (N.P.) (e) – (f) ANON (g) [PHILLIPS (Captain)] (h) – (j) ANON (k) INMAN (H.R.), The Grange, West Heath, Hampstead.	(a) Br macrolep Br microlep (b) Exotic lep Ins (c) Br lep (d) Br lep (e) Ins Min Nh (f) Lep (g) Lep (h) Br lep (i) Exotic lep (j) Cab Sh Birds Nh (k) Br lep	Stevens 897 + 8 bis lots 44pp.	**CH** m.p.s.n.; **GS**; **LC**; **BMNHE**; N s.p.s.n.; SAR
July 10-11	WELCOME (formerly part of ROSEHILL coll.)	Nh (objects from Swiss lake dwellings), 3	Harrods & Allsop 526 lots 44pp + 3 plates	**BMETH**
1941 October 29	BRIGHT (Percy M.)†, Part I	Br butts	Glendining 192 lots 12pp.	**CH** p.n.; **LC**; MAR p.f.n.
1942 January 14	BRIGHT (Percy M.)†, Part II	Br butts	Glendining 225 (incl. 4 blank) + 26 bis lots 12pp.	**CH** p.n.; **GS**; **LC**; MAR p.
February 11	HOPE (Dr. J.)†	Br butts Cab Lib	Glendining 351 lots 20pp.	**CH** p.n.; **GS**; **LC**; MAR m.n. m.p.
March 13	(a) PENNINGTON (F.)† (b) ANON	(a) Br macrolep Lib (b) Br lep	Glendining 177 + 69 bis lots 16pp.	**CH** p.n.; **GS**; **LC**; H; SAR; MAR m.p.m.n.
April 9	(a) BRIGHT (Percy M.)† (b) ANON	(a) Br butts (b) Lib	Debenham & Storr 424 + 1 bis lots 12pp.	**CH** p.m.n.; **GS**; **LC**; MAR m.p. m.n.
May 7	BRIGHT (Percy M.)†	Br butts	Debenham & Storr 324 (incl. 10 blank) lots 8pp.	**CH** p.n.; **GS**; **LC**; MAR p.n.
October 21	CRABTREE (B.H.)	Br butts	Glendining 202 lots 14pp.	**CH** p.n.; **GS**; **LC**; MAR p.n.
December 3	(a) BRIGHT (Percy M.)† (b) [MARCON (Rev. J.N.)] (c) ANON	(a) – (b) Br butts (c) Exotic butts	Debenham & Storr 263 + 1 bis lots 14pp.	**CH** p.m.n.; **GS**; **LC**; MAR p.n.
1943 February 11	(a) STIFF (Phil.)† (b) LEEDS (H.A.) (c) JOY (E.C.), Folkestone† (d) ANON	(a) – (b) Br butts (c) Br butts For butts (d) Butts App Lib	Glendining 260 lots 22pp.	**CH** p.m.n.; **GS**; **LC**; SAR; MAR m.p.m.n.

1943–1945	SOURCE	CONTENTS	AUCTIONEER & SALE CAT.	REFERENCE
April 9	VIPAN (A.M.)	Br macrolep Br microlep	Glendining 282 lots 16pp.	CH m.p.m.n.; GS; LC
October 21	(a) TAIT (R.), Alderley Edge, Cheshire† (b) WHITEHOUSE (Sir Beckwith)† (c) BETHUNE (Major C.P.) (d) TURNER (H.J.)	(a) Br macrolep (b) Exotic lep Cont lep (c) Br lep (d) For lep	Glendining 351 + 4 bis lots 27pp.	CH m.p.s.n.; GS; LC
November 16	WHITEHOUSE (Sir Beckwith)†, Part I[1]	Br macrolep Lib	Glendining 266 + 2 bis lots 22pp.	CH p.m.n.; LC; MAR p.n.
1944 January 25	(a) WHITEHOUSE (Sir Beckwith)†, Part II (b) McLEOD (Sir Murdoch)	(a) Br butts (b) Br butts	Glendining 255 lots 20pp.	CH p.m.n.; LC; MAR p.n.
February 21	(a) WHITEHOUSE (Sir Beckwith),† Part III (b) ARMSTRONG (Major Brian), Ditchling, Sussex.	(a) Br macrolep (b) Br macrolep	Glendining 397 lots 24pp.	CH p.n.; LC; MAR m.p.m.n.
November 2	SHELDON (W.G.)†	Br macrolep Pal macrolep	Glendining 174 lots 11pp.	CH p.m.n.; LC; GS; BMNHE; MAR p. (of butts only)
December 5-6	(a) WHITEHOUSE (Sir Beckworth)†, Part IV (b) EDMONDS (T.H.), Totnes, Devon† (c) MAIN	(a) Br macrolep (moths) (b) Br lep For lep (c) Br macrolep (moths)	Glendining 402 + 1 bis lots 26pp.	CH p.n.; GS; LC
1945 January 18	(a) WOOD (H.), Ashford, Kent† (b) ANON	(a) – (b) Br butts	Glendining 278 lots 18pp.	CH p.n.; GS; LC; SAR; MAR p.n.
January 22-23	SHELDON (W.G.)†	23 i: Entomological lib, 65	Sotheby 535 lots 36pp.	S p.n. copy
February 15	ELLIS (H. Willoughby)†, Part I	Br macrolep	Glendining 250 + 1 bis lots 15pp.	CH p.n.; LC; MAR p.
October 25	(a) ELLIS (H. Willoughby)† Part II (b) SHANNON (D.S.B.) (c) MANNERING (Rev. E.) (d) SMART (Dr. H.D.)† (e) HUDSON (Mrs.)	(a) Br macrolep (b) Br lep For lep (c) – (d) Br lep (e) Exotic ins	Glendining 218 lots 15pp.	CH p.f.n.; LC; MAR
November 16	(a) DEWAR (Dr. D.A.), Durham (b) STIFF (Phil.)† (c) ANON	(a) Br macrolep (b) Br butts (c) Lep	Glendining 182 + 6 bis lots 15pp.	CH p.m.n.; GS; LC; MAR

[1] Lot 94 consisted of two remarkable butterfly varieties—the unique all white (f. *totaalba* C.-H) and the all black (f. *lugens* Oberthür) *Melanargia galathea* L. Vivian Hewitt bought them for £110.00 (the highest price ever paid at auction for two British insects) and so sealed their fate. Sadly, both these historic specimens, as well as other extremely rare butterfly varieties perished from neglect in his outhouse.

1945–1948	SOURCE	CONTENTS	AUCTIONEER & SALE CAT.	REFERENCE
December 17	GREENWOOD (Charles)	Br macrolep (moths)	Glendining 175 lots 12pp.	CH p.m.n.; GS; LC
1946 January 18	(a) BUCKLEY (W.), Cheshire† (b) ANON (c) WARD (Dennis F.), London, N.W.11 (d) ANON	(a) Br macrolep (b) Lib Br lep (c) Br butts (d) Br lep Exotic lep Exotic ins Nh	Glendining 218 + 10 bis lots 18pp.	CH p.m.n., GS; LC; MAR
February 28	(a) SMART (Dr. H.D.)† (b) – (d) ANON	(a) Br lep (b) Ins App (c) App (d) Lep	Glendining 237 lots 16pp.	CH p.m.n.; GS; LC
March 21-22	(a) PENN (W.), Lancing, Sussex† (b) ANON (c) HARTLEY (G.E.), Aberdeen (d) GABBITAS (Miss E.), Ross-on-Wye (e) DANNATT (Walter)†	(a) Br macrolep (b) Br macrolep (moths) (c) Br butts (d) Exotic lep (e) Exotic lep	Glendining 528 + 15 bis lots 47pp.	CH p.m.n.; GS; LC; MAR
November 14	(a) CRABTREE (B.H.), Alderley Edge, Cheshire (b) SCOTT (M.N. Douglas)† (c) EARP (Mrs.) (d) ANON	(a) Br macrolep (b) Lib Br lep Exotic lep App (c) Lep (d) Br lep	Debenham & Storr 445 + 5 bis lots 28pp.	CH p.m.n.; GS; LC; MAR m.p.
December 12	(a) PAGE (H.E.), Blackheath† (b) [COOPER] (c) BIRD† (d) NURSE (Col. C.G.)†	(a) Br lep Cont lep Exotic lep (b) Br lep (c) For lep (d) Lep	Debenham & Storr 345 + 9 bis lots 23pp.	CH p.n., GS; LC
1947 February 6	(a) SMITH (Harold B.)† (b) STIFF (P.)† (c) – (d) ANON	(a) Br macrolep Br microlep (b) Br butts (c) Lib (d) Br macrolep	Debenham & Storr 413 + 7 bis lots 29pp.	CH p.n.; GS; LC; MAR s.p. s.n.
October 22	MARCON (Rev. J.N.), Eastbourne, Sussex, Part I	Br butts	Debenham & Storr 280 + 1 bis lots 21pp.	CH p.n.; GS; LC; MAR p.n.
November 19	(a) HAYWARD (H.C.), Sherborne, Dorset† (b) MARCON (Rev. J.N.) Eastbourne, Sussex, Part II (c) GRANT (J.H.)† (d) GOWING-SCOPES (E.) (e) BIRTLES (H.L.)	(a) Br macrolep (b) Br butts (c) – (e) Lib	Debenham & Storr 301 + 5 bis lots 22pp.	CH p.n.; GS; LC; MAR m.p. m.n.
1948 January 21	(a) JOHNSTONE (J.F.)† (b) GRAVES (P.P.), County Cork (c) REED (Alfred Douglas)† (d) FAWCETT (Dr. H.A.) (e) MARCON (Rev. J.N.), Part III (f) ANON	(a) Br macrolep (b) Br butts (c) For lep (d) For lep (e) Br butts (f) Br butts	Debenham & Storr 324 + 9 bis lots 22pp.	CH p.m.n.; GS; LC; MAR s.p. s.n.
February 18	(a) ANON (b) MARCON (Rev. J.N.), Part IV (c) ANON (d) GRANT (J.H.), Birmingham†	(a) Br lep (b) Br butts (c) For lep (d) Br lep For lep	Debenham & Storr 474 + 44 bis lots 29pp.	CH p.m.n.; GS; LC
November 10	(a) MARCON (Rev. J.N.), Part V (final) (b) – (c) ANON (d) HAWLEY (Col. B.) (e) ANON (f) PATTON-BETHUNE (Major C.L.) (g) – (h) ANON	(a) Br butts (b) Br butts (c) Br moths (d) Br macrolep (e) Exotic ins (f) Exotic butts (g) Lib (h) Lep	Debenham & Storr 383 + 21 bis lots 26pp.	CH p.m.n.; GS; LC; MAR s.p. s.n.

1948–1950	SOURCE	CONTENTS	AUCTIONEER & SALE CAT.	REFERENCE
November 19	(a) TAYLOR (Col. J.E. Campbell) (b) BOWDEN (W.R.) (c) – (d) ANON	(a) Br lep Br ool (18) Irr (b) Br lep Ool (24) (c) Mollusks Nh (5) (d) Lep Cab Ins Nh (8) Irr	Glendining 112 lots 8pp.	LC p.m.n.; GS; CH copy
December 15	ADKIN (B.W.), Pembury, Kent†, Part I	Br butts Lib	Debenham & Storr 428 + 4 bis lots 31pp.	CH p.n.; GS; LC; MAR p.n.
1949 January 19	(a) ADKIN (B.W.), Pembury, Kent†, Part II (b) ANON	(a) Br macrolep (b) Br lep (*Abraxas grossulariata*)	Debenham & Storr 297 + 24 bis lots 21pp.	CH p.n.; GS; LC
March 16	(a) ADKIN (B.W.), Pembury, Kent†, Part III (b) THOMPSON (J. Anthony) (c) ANON (d) GREER (T.), Co. Cork	(a) Br lep (b) – (d) Br butts	Debenham & Storr 497 + 6 bis lots 30pp.	CH p.m.n.; GS LC
October 26	(a) ADKIN (B.W.), Pembury, Kent†, Part IV (b) MAIN (c) THOMPSON (J.A.) (d) CRUTTWELL (G.H.W.) (e) ANON	(a) – (b) Br lep (c) Br butts (d) – (e) Br lep	Debenham & Storr 412 + 8 bis lots 24pp.	CH p.m.n.; GS, LC
December 7	(a) SLATER (P.H.) (b) CHARTRES (S.A.), Eastbourne, Sussex† (c) GREGSON (Col. G.K.) (d) – (g) ANON	(a) Br macrolep (b) Br lep Exotic lep (c) Lib (d) Exotic lep (e) Br butts (f) Br butts (g) Lep	Debenham & Storr 400 + 5 bis lots 21pp.	CH p.m.n.; GS; LC; MAR s.p.
1950 January 25	(a) CLARKE (J.A.), Brockenhurst Hants (b) GREGSON (Col. G.K.) (c) – (e) ANON (f) HUTCHINSON (J. Hely)	(a) Br macrolep (b) Br macrolep For macrolep (c) Br microlep Lib (d) Br macrolep (e) Ins App Cab (f) Br macrolep	Debenham & Storr 420 + 7 bis lots 23pp.	CH p.n., GS; LC
March 15	(a) GREER (T.), Co. Tyrone (b) WATSON (D.)	(a) Br macrolep Lib (b) Br macrolep (moths)	Debenham & Storr 465 + 13 bis lots 29pp.	CH p.m.n.; GS; LC
October 18	(a) FROHAWK (F.W.)† (b) McLEOD (Sir Murdoch)†, Part I (c) ANON	(a) Br butts (b) Br butts Lib App (c) Birds Mam	Debenham & Storr 409 + 3 bis lots 24pp.	CH p.m.n.; GS; LC
November 22	(a) McLEOD (Sir Murdoch)†, Part II (b) WRIGHT (A.E.), Grange† (c) LOWTHER (Dr. R.C.), Grange (d) NORTON (Mrs. R.)	(a) Br butts (b) Br butts Lib (c) Br lep Sh (d) Ool Br lep For lep	Debenham & Storr 273 (incl. 3 blank) + 2 bis lots 18pp.	CH p.m.n.; GS; LC
December 13	(a) McLEOD (Sir Murdoch)†, Part III (b) ANON (c) WRIGHT (A.E.), Grange† (d) LOWTHER (Dr. R.C.), Grange (e) ANON (f) CLASSEY (E.W.)	(a) Br butts (b) Br butts (c) Br macrolep (moths) (d) Br macrolep Lib Exotic butts (e) Lib (f) Br macrolep	Debenham & Storr 357 + 1 bis lots 21pp.	CH p.n.; GS; LC; MAR f.p.

1951–1954	SOURCE	CONTENTS	AUCTIONEER & SALE CAT.	REFERENCE
1951 January 24	(a) EDWARDS (Rev. W.O.W.) (b) RICHARDS (Rev. F.M.) (c) [RUSSELL (S.G.Castle)] (d) ANON (e) SHEPHERD (J.)	(a) Br macrolep (moths) (b) Br lep For lep (c) Br butts (d) Br macrolep (e) Br butts	Debenham & Storr 303 + 8 bis lots 20pp.	CH p.m.n.; GS; LC
March 21	WILLIAMS (Dr. H.B.), Part I	Br macrolep	Debenham & Storr 320 (incl. 3 blank) lots 21pp.	CH p.n.; GS; LC
November 28	HAYNES (H.), Salisbury†	Br macrolep Cab App Lib	Debenham & Storr 351 + 1 bis lots 21pp.	CH p.m.n.; GS; LC
December 19	(a) WILLIAMS (Dr. H.B.), Part II (b) ANTRAM (C.B.) (c) EUSTACE (E.M.)† (d) WATSON (A.) and NEWEY (H.Foster)	(a) – (b) Br butts (c) Br macrolep (d) Br lep For lep	Debenham & Storr 246 (incl. 6 blank) lots 16pp.	CH m.p.m.n.; GS; LC
1952 January 23	(a) HUGHES (C.N.) (b) – (g) ANON	(a) – (b) Br lep (c) Br butts (d) Lep (e) Br butts (f) Br moths (g) Br macrolep	Debenham & Storr 375 + 2 bis lots 23pp.	CH p.n.; GS; LC
November 26 (19 in error)	(a) JAMES (Russell E.) (b) ANON (c) JOHNSTONE (J.) (d) WILLIAMS (Dr. H.B.), Part III (e) – (f) ANON (g) BUXTON (Prof. P.A.) (h) ANON	(a) Br macrolep (b) Lep (*Pieris napi*) (c) – (d) Br butts (e) Br moths (f) Ool (*Aepyornis*) (g) Br macrolep (moths) (h) Br macrolep Br microlep Cab Lib	Debenham & Storr 283 + 50 bis lots 20pp.	CH p.m.n.; GS; LC
1953 January 21	(a) BAYNES (E.S.A.) (b) [NEWMAN (L.H.)] (c) ANON (d) [CUNNINGHAM (Major)] (e) – (f) ANON	(a) Br lep Cont lep (b) Lep (*Abraxas grossulariata*) (c) Br lep Exotic lep (d) Br lep (incl. *Angeronia prunaria*) (e) – (f) Br butts	Debenham & Storr 392 + 31 bis lots 29pp.	CH p.n.; LC p.n.; GS; MAR p.
October 28	(a) BOSTOCK (E.D.), Folkestone† Part I (b) ADAMS (T.W.) (c) DUTTON (J.P.) (d) RUDDY (Mrs.) (e) WILKINSON (W.) (f) GREENWOOD (Charles), Guiting Power	(a) Br butts (b) For butts (c) Br macrolep (d) Br lep (e) Br lep For lep (f) Br butts	Debenham & Storr 579 + 7 bis lots 28pp.	CH p.n.; LC; GS
December 9	(a) BOSTOCK (E.D.), Folkestone† Part II (b) ANON (c) SHARMAN (F.W.) (d) HUMPHRY-BAKER (Mrs. D.) (e) ANON	(a) Br macrolep (moths) (b) Lep (*Abraxas grossulariata*) (c) Br macrolep (d) Exotic butts (e) Br lep	Debenham & Storr 330 lots 17pp.	CH p.n.; LC; GS
1954 October 28	(a) WHITE (Dr. E. Barton)† (b) ANON	(a) Br macrolep Lib (b) Br macrolep	Debenham & Storr 394 + 11 bis lots 21pp.	CH p.m.n.; GS

1955–1959	SOURCE	CONTENTS	AUCTIONEER & SALE CAT.	REFERENCE
1955 February 10	(a) RUSSELL (G.M.) (b) ANON (c) [NEWMAN (L.H.)] (d) – (g) ANON	(a) Br butts Exotic butts (b) – (e) Br butts (f) Exotic col (g) Br macrolep	Debenham & Storr 299 + 4 bis lots 19pp.	CH p.n.; GS; LC
June 8	ANON, Bowerholme, Shrubbs Hill, Lyndhurst, Hampshire	Ins Cab App, 41	Sawbridge, *house sale* 465 (incl. 34 blank) lots 10pp.	CH; LC
November 9	RUSSELL (S.G. Castle)	Entomological lib, 17	Sotheby 28 lots (819–846) 4pp. (105-108)	LC p.n. (of rel. lots only)
November 24	(a) RUSSELL (S.G. Castle) (b) BESSEMER (H. Douglas) (c) – (d) ANON (e) DOUDNEY (S.P.)† (f) – (g) ANON	(a) Br butts Exotic butts (b) Br macrolep (c) Br butts (d) Br macrolep (moths) (e) – (f) Br macrolep (g) Cab	Debenham & Storr 364 + 59 bis lots 29pp.	CH p.n.; GS
1957 December 12	(a) KING (E. Bolton) (b) COLMAN (K.E.S.) (c) STIFF (P.A.)† (d) – (f) ANON (g) CLUTTERBUCK (C.Granville), Gloucester†	(a) Br butts Cont butts (b) – (c) Br butts (d) Br moths Exotic col (e) Exotic butts (f) Br butts (g) Br macrolep Br microlep Lib App	Debenham & Storr 427 + 4 bis lots 24pp.	CH p.m.n.; GS
1958 April 10	(a) WILLSHEE (C.J.) (b) ANON (c) HAMMOND (L.F.) (d) TODD (R.G.) (e) SMITH (S. Gordon), Part I (f) ELLIS (G.)	(a) Br lep Cont lep (b) Br macrolep (c) Br butts (d) Br macrolep (e) Br butts (f) Br macrolep	Debenham & Storr 348 + 1 bis lots 17pp.	CH p.m.n.; GS
September 17-19	MALCOLM (Col. George), Poltalloch House, Kilmartin, Lochgilphead, Argyll	18 ix: Birds, 7	Edmiston, *house sale* 1179 lots 29pp.	CH p. copy
October 23	(a) GRAHAM (Eric W.) (b) ANON (c) ROBSON (J.P.) (d) JONES (Arthur) (e) WINSTON (Dr. G.D.)† (f) LOBB (J.) (g) ANON (h) BROWN (S.A.)† (i) MOUNSEY (John D.) (j) ANON (k) SMITH (S. Gordon), Part II	(a) – (c) Lib (d) Exotic lep Lib App (e) Br macrolep Cab Lib (f) Br lep For lep (g) Br lep (h) – (j) Br macrolep (k) Br butts	Debenham & Storr 426 + 50 bis lots 28pp.	CH p.n.; GS
1959 February 26	(a) OLIVER (G.B.) (b) ANON (c) SMITH (S. Gordon), Part III	(a) – (b) Br butts (c) Br macrolep (moths)	Debenham & Storr 387 + 3 bis lots 20pp.	CH m.p.f.n.; GS
October 29	(a) BOWATER (Col. W.) (b) BIRCH (L.) (c) SUTTON (G.P.)	(a) Br lep Cont lep Exotic lep (b) Br butts Br macrolep (*Arctia caja*) (c) Br macrolep	Debenham & Storr 415 + 29 bis lots 27pp. + 1 plate	CH p.; GS

1961–1967	SOURCE	CONTENTS	AUCTIONEER & SALE CAT.	REFERENCE
1961 February 23	(a) QUIBELL (William)† (b) – (c) ANON	(a) Br macrolep (b) – (c) Br butts	Debenham & Storr 252 + 39 bis lots 19pp.	**CH** f.p.f.n.
November 9	MICHAUD (Dr. J.)†	Br lep Cont lep Exotic lep Ins Cab App	Debenham & Storr 451 + 2 bis lots 31pp.	**CH** m.p.m.n.
1962 January 25	(a) BURTON (P.) (b) – (c) ANON	(a) Br butts (b) Lib (c) Hym App	Debenham & Storr 398 + 6 bis lots 32pp.	**LC** p.n.; **CH** p.n. copy
March 1	(a) BURTON (P.) (b) – (d) ANON (e) SMITH (S. Gordon), Part IV (final)	(a) Br macrolep (b) Br butts (c) Br lep App (d) Microlep Col Hem Orth Hym Exotic hym (e) Br macrolep	Debenham & Storr 400 + 4 bis lots 31pp.	**CH** s.p.s.n.
October 25	WATKINSON (Canon George)†, Part I	Br lep Cont lep Exotic lep	Debenham & Storr 400 + 4 bis lots 30pp.	**CH**
November 29	WATKINSON (Canon George)†, Part II	Br macrolep	Debenham & Storr 418 + 4 bis lots 23pp.	**CH; LC; GS**
1963 November 4-5	'A Lady'	Lib	Sotheby 437 lots 85pp.	**BMNHZ**
November 14	(a) VALENTINE (Arthur)† (b) ANON (c) SOWELS (Frank), Thetford	(a) – (b) Br butts (c) Br lep Exotic lep	Debenham & Storr 329 lots 16pp.	**CH** p.m.n.
November 14-15	JOHNSON (Miss L.E.M.), Rose Cottage, Smalldale, Bradwell, Derbyshire†[1]	14 xi: Min Fos Sh Lep Col Birds Cab, 8	Spencer, *house sale* 925 lots 11pp.	**BMNHL** copy
1965 October 28	(a) WOOLLETT (G.F.C.)† (b) LEIVERS (c) FRASER	(a) Br butts Exotic butts Cont butts (b) – (c) Br butts	Debenham & Storr 311 + 1 bis lots 16pp.	**CH** p.m.n.
1967 August 30-31	FLATTERS & GARNETT, Biologists & Scientific Instrument Makers, Manchester	Ins Bot Zoological specimens Nh App, 329	Singleton, *Manchester* 880 lots 72pp.	**CH**

[1] The chief feature of interest was the geological collections. These were in four 24 drawer cabinets (lots 471–474) and a show case (lot 469). Doncaster Mus. bought lots 469, 471, 473 and 474.

1967–1970	SOURCE	CONTENTS	AUCTIONEER & SALE CAT.	REFERENCE
November 10	ROWBERRY (D.)[1]	Br lep For lep	Debenham & Storr 293 lots 16pp.	CH p.m.n.
1968 July 9	(a) NORTHUMBERLAND (Duke of) (b) ANON (c) USHAW COLLEGE, Durham (d) CLARKE (Sir Ralph S.) (e) DRURY-LOWE (Capt. P.J.B.) (f) DAY (J. Wentworth) (g) ANON	(a) Min Ores (b) Min Cor Butts Nh (c) Sh Birds (d) – (f) Birds (g) Cab	Sotheby 160 lots 34pp. + 4 plates	CH p.n.
December 5	(a) DAY (J. Wentworth) (b) ANON (c) STORER (G.H.)† (d) ANON (e) NORTHUMBERLAND (Duke of) (f) ANON (g) DRURY-LOWE (Capt. P.J.B.), Locko Park (h) ANON (i) TURNER-BRIDGER (R.C.) (j) ANON	(a) – (b) Birds (c) Birds For lep (d) Birds Lep (e) Sh Min (f) Sh Cor Nh (g) Min (h) Min Fos Ores (i) Br lep (j) Lep Birds Min	Sotheby 164 lots 31 + 2pp. + 3 plates	CH p.n.
1969 June 10	ANON	Min Fos Met Ores	Sotheby 176 + 1 bis lots 31pp. + 5 plates	CH
July 31	ANON	Exotic butts Ins Sh Birds Min Fos Ores Cor Nh	Sotheby 224 lots 33pp.	CH
November 6	BACK (Major D.H.L.), Hethersett Hall, near Norwich, Norfolk[2]	Ool Sh Min Fos Lep Cab Birds, 24	Hall & Palmer, *house sale* 580 + 1 bis lots 16pp.	CH
December 22	(a) ANON (b) 'WARWICK CASTLE RESETTLEMENT' (c) ANON (d) DORSET COUNTY MUSEUM (e) ANON	(a) Birds Lep Ins (b) Min Fos (c) Min Ores Fos (d) Sh (e) Sh Cor Fish Nh	Sotheby 230 + 8 bis lots 36 + 4pp. + 2 plates	CH p.n.
1970 March 5	(a) ANON (b) 'A Gentleman' (c) – (e) ANON (f) GRANT INSTITUTE OF GEOLOGY, EDINBURGH (g) ANON (h) DORSET COUNTY MUSEUM (i) ANON	(a) – (b) Min (c) Min Ores Met (d) Min (e) Min Fos (f) Fos (*Ichthyosaur*)[3] (g) – (i) Sh	Sotheby 374 + 8 bis lots 59pp. + 16 plates (incl. 1 coloured)	CH
April 23	ANON [Although advertised, this sale never took place]	Lep	Debenham & Storr	*Ent. Gaz.*, **21** : 62
July 2	(a) ANON (b) CURWEN (H.C.), F.G.S. (c) – (d) ANON (e) ROSA (Dr. Lewis G.) (f) ANON	(a) Min Fos (b) Min (c) Sh (d) Lep Col (e) Exotic butts Cab (f) Birds	Sotheby 225 + 7 bis lots 36 + 4pp.	CH p.n.

[1] This coll. contained little of interest, but the sale was unusual in that it was catalogued by the vendor.

[2] The Royal Society for the Preservation of Birds summoned the auctioneer for selling 13 eggs, all of which were *at least* 50 years old. The magistrate said: 'It seems doubtful whether a case like this has ever been heard before in this country.' The defendant was granted a conditional discharge.

[3] A fine specimen of *Ichthyosaurus platyodon* (Conybeare) from the Lower Lias, Lyme Regis, Dorset was bought by Martin Bodmer, of Switzerland, for £3,200.

1970–1971	SOURCE	CONTENTS	AUCTIONEER & SALE CAT.	REFERENCE
October 15	ANON	Min	Sotheby 180 lots 2 + 47pp. (incl. 12 plates)	CH p.n.
November 3-5	ODD (D.A.)†, ANON	5 xi: Lib Birds Mam Sh Ins Fos Min Cab Nh, 234	King & Chasemore, *Pulborough* 1570 lots 32pp.	CH
December 7	(a) ANON (b) COLLORD (James) (c) – (d) ANON (e) EVANS (Mrs. Jeanne) (f) ANON (g) 'A Gentleman'	(a) Min (b) Met (c) Min (d) Fos (e) Fos (*Ichthyosaurus*) (f) Fos (*Oreodon*) (g) Fos (*Ichthyosaur*, 2)[1]	Sotheby 174 lots 4 + 35pp. (incl. 8 plates)	CH p.n.
1971 March 4	(a) ANON (b) CURWEN (H.C.) (c) ANON (d) HOPE (Mrs. A.E.) (e) ANON (f) RICHARDSON (J.) (g) ANON (h) WARD (Dennis) (i) ANON (j) 'A Gentleman' (k) – (l) ANON (m) GERRARD (n) ANON (o) RABEN-LEVETZAU (Baron), Aalholm Castle, Denmark	(a) – (e) Min (f) Min (John RUSKIN coll.) (g) Birds Exotic lep Sh Nh (h) Br lep For lep (i) Fos (j) Ool (*Aepyornis*) (k) Sh (l) – (n) Birds (o) Great Auk	Sotheby 260 lots 4 + 47pp. (incl. 7 plates)	CH p.n.
March 16	ANON	Nh, 1	Bearnes, *Torquay* 17 + 1 bis lots 24pp.	CH
March 30	ANON	Sh Min Fos Ool Nh, 16	Bearnes, *Torquay* 622 (incl. 43 blank) + 1 bis lots 56pp. + 6 plates	CH
June 14-17, 21	McCARDLE (L.F.), Sheffield Park, near Uckfield, Sussex†	16 vi: Birds Nh, 38	St.John Smith, Graves & Pilcher *house sale* 2310 (incl. 113 blank) + 8 bis lots 130pp. + 16 plates	CH
June 23	ANON	Birds Ool, 4	Hall Wateridge & Owen, *Welsh Bridge* 393 (incl. 26 blank) + 1 bis lots 8pp.	CH s.p.s.n.

[1] The second example in this sale of the ichthyosaur *Eurypterygnis dorsetensis* was an almost complete specimen found at Lyme Regis in 1969. Cleveland Natural History Museum bought it for £1,050.

1971–1972	SOURCE	CONTENTS	AUCTIONEER & SALE CAT.	REFERENCE
July 15	ANON	Min Fos Sh	Sotheby 372 + 10 bis lots 6 + 48pp. + 11 plates	**CH** p.n.
October 19-21	PARKER (Sir William), Llangattock Court, Crickhowell, Breconshire†	19 x: Butts Moths Cab Birds, 10	Bruton, Knowles, *house sale* 846 lots 48pp.	**CH** p.
December 13	ANON	Min Fos Met	Sotheby 246 lots 4 + 26pp.	**CH** p.n.
1972 February 8	TENNANT (Prof. J.)†	Min Sh Fos Ool Cab, 24	Stacey, *Leigh-on-Sea, Essex* 261 + 82 bis lots 8pp.	**BMNHL** p.n. copy
February 17-18	McDONALD (R.J.), Member of the Jourdain Soc. & B.T.O.	18 ii: Ool Cab Card index Lib, 4	Button, Menhenitt & Mutton, *Wadebridge, Cornwall* 1473 lots 45pp. + 4 plates	**CH** p.
February 23	(a) [LEACH] (b) – (c) ANON	(a) Sh (b) Fos Fish (c) Birds Lep Nh	Spencer, *Retford, Notts* 369 + 2 bis lots 24pp.	**CH** p.; **NMW**
March 17	(a) ANON (b) NELSON'S PHARMACY, Duke Street (c) ABINGDON MUSEUM (d) ANON (e) TELFORD (A.E.), incl. William BUCKLEY, F.Z.S., Skelmanthorpe, Huddersfield (f) BANKFIELD MUSEUM, Halifax, Yorkshire	(a) Min Met Fos (b) Ool (*Aepyornis*) (c) Fos (incl. *Ichthyosaur*) (d) Min Fos Sh Exotic butts Birds Cor Nh (e) Br lep (f) Birds (Passenger pigeons)	Sotheby 416 + 5 bis lots 6 + 57pp. + 58 illustrations	**CH** p.n.
March 20	STEVENS (O.C.), Leigh Beck, Canvey Island, Essex†	Butts Exotic ins Catalogue Lib, 24	Stacey, *Leigh-on-Sea, Essex* 261 + 82 bis lots 8pp.	**BMNHL** p.n. copy
March 22	ANON	Birds Mam, 6	McCartney, Morris, Barker, *Ludlow, Salop*	**CH** p. copy (inc.)
April 26	ANON	Birds Sh Fos Min Lep Col Fish Mam Cab Nh	Messenger, May, Baverstock, *Godalming, Surrey*. 252 lots 24pp.	**CH** p.

1972	SOURCE	CONTENTS	AUCTIONEER & SALE CAT.	REFERENCE
May 17	[STRICKLAND (Colin)], Lees Court, Sheldwich, near Faversham, Kent.	Birds Nh, 85	Burrows, *house sale* 510 (incl. 69 blank) lots 18pp.	CH p.n. (of bird lots only)
May 17	BLACKWELL-WOOD (Neville)	Lep Ins Nh, 48	Sheppard, *South Normanton, Derbyshire* 337 + 22 bis lots 24pp.	CH m.p.m.n. (of rel. lots only)
June 9	CRANE (Miss E.G.), Great Bealings, near Woodbridge, Suffolk†	Birds, 14	Spear, *house sale* 460 lots 17pp. + 8 illustrations	CH p. (of rel. lots only)
July 5	ANON	Birds Nh, 25	Wooley & Wallis, *Salisbury* 572 (incl. 32 blank) + 6 bis lots 26pp.	CH p. (of rel. lots only)
July 12	(a) PARLETT (Justice) (b) – (c) ANON	(a) Exotic butts (b) Sh (c) Min Fos Nh	Sotheby 212 + 3 bis lots 4 + 19pp. incl. 7 illustrations	CH p.n.
August 8	ANON	Birds Mam Rept Ool Fish	Button, Menhenett & Mutton, *Wadebridge, Cornwall* 78 lots 2pp.	CH copy
October 11-12	[LE BLOND, natural history dealers]	Birds Mam Ool Rept Fish Lep Ins Sh Fos Nh	Stocker & Roberts, *Hampstead* 738 + 82 + 7 bis lots 27 + 11pp.	CH
November 14-16	ANON	Min, 9	Lane, *Penzance, Cornwall* 2pp.	CH p. copy (inc.)
November 27	ANON	Sh Min Met Birds Fos Cab Exotic butts Fish Ores	Sotheby 309 + 4 bis lots 6 + 29pp. incl. 11 plates	CH p.n.
November 30	KEMPTON MANOR HOTEL, Hothfield, near Ashford, Kent	Birds HH, 16	Burrows, *house sale* 607 (incl. 22 blank) + 1 bis lots 31pp.	CH

1972	SOURCE	CONTENTS	AUCTIONEER & SALE CAT.	REFERENCE
December 12-15	ANON	15 xii: Min Fos Ins Cab Mam HH Birds Nh, 192	King & Chasemore, *Pulborough* 2312 (incl. 11 blank) + 63 bis lots 72pp.	CH

SALES WITHOUT INDICATION OF THE YEAR OR DATE

(Arranged alphabetically under the names of the vendors or those who formed the collections.)

	SOURCE	CONTENTS	AUCTIONEER & SALE CAT.	REFERENCE
Early 19th cent.	[ANON]	Birds Ins	[? King & Lochée] 122 + 3 bis lots 8pp.	**BRAD** inc. (defective title-page)
	BLACKETT (Charles Powell)	Nh		Allingham (1924:36)
	BRAND	Sh		Allingham (1924:24)
?1859	BREE (C.N.) [possibly not a sale by auction]	Fos		Advert. on cover of *The Geologist*, August 1859
18th cent.	CHURCH, apothecary	Ins	Paterson	Da Costa (1812:515)
Early 19th cent.	DAVIES (General Thomas), Royal Artillery [possibly not a sale by auction]	Ins Birds		Fletcher, 1920, *Proc. Linn. Soc. N.S.W.*, **45**:575
[?1774] May 26	GAMBERINI	Sh Min Ores Herb Nh, 18	Cock [178 lots] [8pp.]	**BML** inc.
c. 1797	GREVILLE (Rt. Hon. Charles Francis), incl. Baron Ignaz von BORN [b. 1742 d. 1791]	Min	King & Lochee	Bournon, *Cat. Coll. Min.* (1813), p.xxxvi
[1773]	GUY (Richard)	Nh	Paterson	Da Costa (1812:206)
	HAYES (Dr.)	Min		Allingham (1924:25)
c. 1911	HIGGINS (E.T.)	Fos	Stevens	Sherborn (1940:69)
	HITCHMAN	Arboretum	Stevens, *Leamington* 158pp.	Allingham (1924:89)
	HOYER (Jacob)	Ins		Allingham (1924:33)
	INWOOD	Fos		Allingham (1924:36)
[? 1802] April 5 ffdd. (12 days)	KEATE (George)	Sh Cor Min Ores Ins Nh	King 1440 lots 67pp.	HEH; **BMNHL** copy; **BMNHZ**
	KILCOURSIE (Lord)	Sh		Allingham (1924:24)
	KENT	Min		Allingham (1924:25)
	MARSHALL (Edward)	Br lep		Horn & Kahle (1936:166)

SALES WITHOUT INDICATION OF THE YEAR OR DATE

(Arranged alphabetically under the names of the vendors or those who formed the collections.)

	SOURCE	CONTENTS	AUCTIONEER & SALE CAT.	REFERENCE
	MENGE	Min		Allingham (1924:25)
	MIDGLEY	Min		Allingham (1924:25)
	PARKE (S.)	Min		Allingham (1924:25)
'13 or 14 days' sale'	SEYMER (Henry), F.L.S. [b. 1745 d. 1800]	Sh Ins Fos Nh	Hutchins	MS. note by J.C.Dale in the Hope Dept. copy of the Willett sale cat. of 6.xii. 1813 ffdd.
	SOLANDER (D.C.) [b. 1736 d. 1782]	[Fos]		Cited in a ms. cat. of a personal fossil coll. [of Charles Bennett, 9th EARL OF TANKERVILLE] in Geological Survey Mus., Exhibition Rd., London (*teste* H.Torrens, *in litt.*)
'eight days' sale'	STRANGE (John)	Nh		Allingham (1924:20)
c. 1842	SWAFFHAM PRIOR NATURAL HISTORY SOCIETY, incl. JERMYN family collections	Nh		Xerox of unpublished ms. note by Edmond Jermyn in possession of Miss Sheila Hynes (*teste* C.MacKechnie Jarvis *in litt.*)

Index

The index gives the names of the collections, the names of the vendors, or the names of those who formed the collections.

A

Abbot, Rev.W.H., 138
Abbott, P.W., 137
Abbott, S., 160
Abbott, W.J.L., 165
Abingdon Museum, 176
Adams, F.C., 12, 156
Adams, H.J., 152, 153
Adams, T.W., 171
Adamson, J., 47, 90
Adkin, B.W., 170(4)
Adkin, R., 165
Aiton, J.T., 92
Allen, J.E.R., 155
Allen, L.F., 146
Allman, G.J., 142
Ames, J., 58
Amis..., J. da, 95(2)
Anderson, E., 119
Anderson, J., 165
Angus, Mrs., 78
Anning, Miss M., 78
Anstice, R., 88
Anstruther, 123
Antram, C.B., 171
Ardesoif, J., 63
Argent, 87
Armfield, 100, 102
Armitage, E., 150
Armstrong, 100
Armstrong, Maj.B., 168
Arnold, G., 151, 152
Ash, Rev.C.D., 146
Ashby, R., 133
Ashmead, G.B., 98
Ashmolean Museum, 136
Askew, H.W., 109
Astley, F.D., 77
Atherton, J., 148
Atkin, G., 101
Atkinson, E.T., 131
Atkinson, W., 138
Atkinson, W.S., 118
Auckland, J.T., 101
Auld, H.A., 147
Aylesford, Countess of, 40, 41, 83(3), 84, 85, 86

B

Babington, C., 121, 123
Back, Maj.D.H.L., 174
Backhouse, F.W., 120
Backhouse, J., 146
Bacon, 96
Bacon, Rev.T., 108
Bagot, 111
Baikie, 103
Baily, W.S., 164
Baker, 58
Baker, G., 87
Baker, H., 3, 60
Baker, J., 97
Baker, Lady, 137
Bakewell, R., 88
Baldock, G.R., 166
Balston, W.E., 156
Baly, 124
Bankes, E.R., 163
Bankes, Sir J., 72
Bankfield Museum, 176
Banks, D., 89
Banks, Miss, 134
Baracé, Comte de, 121
Barclay, Sir D., 49, 50, 125, 134
Barclay, Col.H., 150(4)
Barford, H., 72
Barker, H.W., 149
Barker, Sir R., 64
Barlow (see Barton,Capt.)
Barlow, F., 112
Barnes, W., 80
Barraud, P.J., 156
Barrett, C.G., 11, 144(2), 147(2)
Barrington, Mrs., 58
Barron, C., 130
Barry, M., 101, 113
Bartlett, 145, 149(2)
Bartlett, A.D., 143
Bartlett, F., 131
Bartlett, Rev.T.O., 86
Barton, Capt., 120
Barton, S., 135
Bates, 121
Battley, A.U., 157
Bauer, F., 86
Baxter, J., 162
Baxter, T., 153, 164
Baylee, Rev.J., 115
Bayly, G., 66
Baynes, E.S.A., 171
Baynes, Sir W.J.W., 134
Bazett, Mrs., 144

Beauclerk, T., 61
Beaumont, A., 11, 144(2), 145, 147
Beckford, W., 146
Beckwith, 6
Beddell, 112
Beecher, H.M., 129
Belcher, Adm.Sir E., 111
Bell, 101
Bell, J., 114
Bell, T., 18, 113
Bellamy, C., 72
Bennett, J.W., 17
Bennett, W., 124
Bensher, 114
Bentinck, Count, 75
Bentley, W., 101
Benza, 29
Beresford, J.C., 72
Bernard, C., 57
Bertling, A., 145, 149(2)
Bessemer, H.D., 172
Bethell, Capt.R., 157
Bethune, Maj.C.L.P., 168, 169
Beveridge, Mrs., 124
Bewley, 110
Bibbs, 142
Bidwell, E., 117(2), 140, 141(2), 145
Billups, T.R., 12, 157
Birch Col., 36, 78
Birch, L., 172
Birchall, E., 114
Bircham, F.S., 116
Bird, 169
Bird, J., 121(2)
Birmingham Phil.Inst., 93
Birtles, 169
Bjorkborn, 146
Blaber, J., 123(2)
Blackall, W., 120
Blackburn, Rev.T., 107
Blackett, C.P., 179
Blackmore, T., 107, 110, 111
Blackwell-Wood, N., 177
Bladen, W.W., 121, 137
Blair, A., 149
Bland, M., 92
Blanckley, 58
Blenkarn, S.A., 153
Bligh, Mrs., 47, 79

181

N.H.A.—M

Bliss, 123
Boerhaave, 57
Bolton, G., 147
Bond, B., 63
Bond, F., 12, 159
Bonhote, J.L., 158
Bonelli, A., 77
Booth, 152
Booth, H.T., 148, 160
Born, Baron I.von, 179
Bosc de la Calmette, Col., 17, 60
Bostock, E.D., 171(2)
Botanical Soc., 97
Boucard, 61, 63, 116, 127, 128
Bouchard, P., 103
Bouck, Baron, 166(2)
Bourbon, Duke de, 75
Bousfield, W.S., 100
Bowater, Col.W., 172
Bowden, W.R., 170
Bowen-Robertson, Maj.R., 157
Bower, B.A., 158
Bowerbank, J.S., 103
Bowles, W., 111
Boys, Maj., 90
Boys, Capt.W.J.E., 91(2)
Braam, A.E.van, 66
Brackenbury, Rev.E.B., 138
Brady, L.S., 157
Brand, 179
Brauer, R., 144
Bree, C.N., 179
Bree, M., 71
Brewer, J.A., 107
Brewin, S., 100
Bridger, 114
Briggs, C.A., 132(3)
Briggs, T.H., 9, 155
Bright, J., 100
Bright, P.M., 13, 136(2), 156, 166, 167(5)
British Museum, 68, 76
Broad, C., 133
Brockholes, J., 149
Brocklehurst, W.S., 166
Broderip, W., 78
Broderip, W.J., 99
Brookes, J., 16, 80, 81(2)
Broughton Castle, 85
Broughton, E., 100
Brown, E., 8, 111
Brown, E.W., 161, 165
Brown, G.D., 62
Brown, J.A., 85
Brown, Rev.J.L., 109
Brown, R., 99(2)
Brown, S.A., 172
Brownell, G., 42
Brunton, J., 73
Bryant, 157
Bryson, A., 106
Buckell, E., 153
Buckingham, Countess of, 76
Buckingham, Duke of (see Grenville, R.)

Buckland, Dean, 96, 97
Buckley, G.G., 161
Buckley, W., 169, 176
Bucknill, J.A., 141
Bullen, Rev.A., 150
Bullock, 24, 77(2), 78(3), 79, 80
Bulow, C., 50, 153, 159
Bunbury, Capt., 155(2)
Bunyard, P.F., 144, 163, 166(2)
Burchell, W.J., 103
Burgess, S., 100
Burgon, J.T., 102
Burman, R., 100
Burnell, E., 126
Burney, Rev.H., 9, 127(3), 128
Burrough, Rev.H., 59
Burton, J., 84
Burton, P., 64
Burton, P., 173(2)
Bute, Earl of, 64, 65, 67
Butler, W.E., 161
Butts, T., 88
Buxton, P.A., 171

C
Cairns, R., 151
Caldecot, 58
Caley, G., 81
Calonne, C.A.de, 67, 68, 78
Calvert, J., 133(2), 134
Cameron, E.S., 125
Campbell, Sir J.W.P., 143(2)
Cannon, A.E., 148
Cansdale, F.E., 160
Cantor, 106
Cape of Good Hope, 16, 85
Capper, C., 154
Carbonell, J., 155
Cardew, Col.P.A., 155
Carnegie, D.J., 160
Carter, S., 105
Cassal, 152
Cavendish, F., 73
Challenger Expedition, 121
Chambers, C., 120
Champ, H., 161
Champley, R., 133
Chandler, Rev.J.B., 139
Chant, J., 6, 106
Chapman, T.A., 158
Chappell, J., 148
Charters, S.A., 170
Chaumette, A.de la, 108
Chichester (Marquis of Donegal), 67(2)
Children, J.G., 7, 86
Cholmondeley, R., 119, 132, 133(2)
Christian of Waldeck, Princess, 85
Church, 4, 179
Claremont, C.C., 140
Clark, Rev.H., 103, 105
Clark, J.A., 11, 149(4), 151, 170
Clark, S.V., 128, 159

Clarke, Sir R.E., 174
Clarke, R.M., 101
Clark-Kennedy, Capt.A.W.M., 119(2)
Classey, E.W., 170
Claxton, Rev.W., 149
Clerk, J., 82
Clutterbuck, C.G., 161, 172
Cobbe, Col., 84
Cockburn, C., 128
Cockburn, Lord, 95
Cohen, 69
Cole, J.W., 93
Colebrooke, J., 60
Coles, H., 105
Collier, J., 165
Collins, H., 122
Collord, J., 175
Colman, K.E.S., 172
Coltart, N.B., 164
Condamine, Rev.H.de la, 94
Connel, 107
Connop, Mrs.A., 134
Conquest, G.H., 152
Constable, W., 64
Cooke, 134
Cooke, Capt., 46, 47, 72
Cooke, B., 9, 116
Cooke & Son, 126
Cooke, O.F.E., 156
Coombe, 75
Cooper, 169
Cooper, A., 106
Cooper, B., 126, 163
Cooper, E., 126
Cooper, J.A., 132(2)
Cooper, S., 158
Copeland, A.T., 152
Corbett, B., 104
Corbett, J., 144
Corder, J.W., 166
Cornell, E., 163
Cornford, Rev.E.B., 156
Cottam, A., 151
Cotton, Rev.H.S., 90
Cotton, T.A., 157
Cox, C.S.B., 150
Cox, G.S., 125
Cox, J.C., 50, 142(2)
Coxon, H., 162
Crabtree, B.H., 165(2), 167, 169
Craft, J.G., 164
Cramer, C., 111
Crane, Miss E.G., 177
Cranch, 76
Craven, A.E., 124
Cregoe, J.P., 133
Crewe, Sir V.H., 160(3), 161(3)
Crichton, Sir A., 43, 80
Crichton, A.W., 115
Cripps, 107
Crisp, 116
Crombie, 148
Crompton, S., 137
Cross, F.B., 162

Cross, W.J., 147
Crotch, G.R., 135
Crotch, W.D., 136
Crouch, W., 129
Crowfoot, 164
Crowfoot, W.M., 161
Crowley, P., 137, 138(3), 139(3)
Cruttwell, Canon C.T., 159
Cruttwell, G.H.W., 170
Cullen, 111
Cuming, H., 47, 48, 103, 104
Cumming, R.G., 104(2)
Cunningham, Maj., 171
Curtis, J., 6, 102(2), 131
Curwen, H.C., 174, 175
Cutting, 92

D

Dacie, J.C., 162
Da Costa, S.J., 50, 139, 146
Dale, C.W., 145
Dalgleish, J.J., 162(3)
Dalglish, A.A., 160
Damon, R.F., 149, 150, 162
Dannatt, W., 169
Daubenay, G.M., 114
Davall, E., 68
Davenport, H.S., 129, 132
David-Ward, J., 53
Davies, G., 92
Davies, Gen.T., 179
Davis, J.W., 146
Davis, M., 122
Davis, T.C., 72
Davison, W., 114
Day, Rev.A., 158
Day, G.O., 144, 147
Day, J.W., 174(2)
Delafons, J.P., 83
Demancha, J., 154
D'Emmich, G., 118
Dennison, J., 48, 49, 103
Derby, Earl of (see Smith,E.)
Deseglisé, 130
Desvignes, T., 8, 106
Deuchars, 107
Devisme, W., 61
Dewar, D.A., 168
Dewey, 165
Dick, P., 78
Dicksee, 155(2)
Dicksee, A., 126, 127, 164
Digby, C.R., 119
Dix, Rev.J., 111
Dixon, Rear Adm.M.H., 99
Dobrée-Fox, Rev., 145
Doeg, T.E., 161
Doncaster, 134, 135
Donegal, Marquis of (see Chichester)
D'Onis, Chevalier, 79
Donisthorp, 124
Donovan, E., 4, 5, 16, 77
Dormer, Lord, 137
Dorset County Museum, 174(2)

Doubleday, E., 91
Doubleday, H., 107, 108, 110
Doudney, S.P., 172
Douglas, Lady, 63(2)
Douglas-Fox, J., 139
Downes, Rev.A.M., 160
Downing, J.W., 129(2)
Drake, 70
Dree, Marquis de, 79, 80
Dresser, H.E., 105
Drury, D., 4, 69
Drury-Lowe, Capt.P.J.B., 174(2)
Du Chaillu, 17, 102
Duckworth, H., 158
Dudgeon, 124
Duff, R., 105, 106
Dunn, N., 92
Dunston, J., 90
Durant, Lt.Col., 89
Durrant, H., 126
Dutton, J.P., 171
Dyer, S., 59

E

Earp, Mrs., 169
Eastwood, J.E., 163
Eddrup, Rev.T.B., 158
Edington, 94
Edkins, W., 124
Edmonds, A., 107
Edmonds, T.H., 168
Edwards, 97
Edwards, 110
Edwards, Miss A.D., 163
Edwards, Rev.W.O.W., 171
Eedle, T., 121
Elisha, G., 10, 134(2)
Ellis, G., 172
Ellis, H.W., 168(2)
Ellis, J., 15, 64
Elwes, H.J., 159, 160
Emsley, F., 146
Enderby, Mrs. 90
Engleheart, 119
Engleheart, N.B., 107
Entomological Club, 7, 87, 161
Entomological Society of London, Royal, 7, 98, 102
Escombe, 127
Esson, 159
Estridge, H.W., 138
Etheridge, 143
Eton College Museum, 135
Eustace, E.M., 171
Evans, 122, 143
Evans, Mrs. J., 175
Evans, H.N., 102
Evans, W.F., 107, 110
Ewer, S., 71

F

F......, R.T., 102
Falconar, M., 61
Farn, A.B., 158(5)
Farr, W.B., 127

Farre, 114
Farren, W., 10, 131(2)
Farwell, C.G., 159
Fawcett, Col., 160
Fawcett, H.A., 169
Featherstonhaugh, G.W., 105
Felder, 113
Fenn, C., 107, 161
Fenn, J., 107
Fenton, I., 73
Field, 71
Field, 118
Field, L., 124, 129, 130, 133, 135
Field-Fisher, T., 137
Fielden, Col., 157
Finch, Lady, 100
Finlay, G., 111
Finnell, Capt.B., 84
Fisher, Miss, 93
Fisher, Rev.F.H., 150
Fitton, 96
Fitzroy, Lady, 63
Flatters & Garnett, 173
Fleming, 107
Fleming, G., 59
Fleming, J.McA., 163
Fludyer, Sir T., 59
Folkes, W., 94
Folliott, G., 118
Footit, W.F., 115(3), 116(2)
Forbes, J., 125
Forster, E., 91
Forster, I.(see Foster)
Forster, J., 42, 71(5)
Forster, W., 161
Forsyth, W., 83
Fortune, R., 108
Foster (or Forster), I., 42, 62(3)
Fox, N.P., 167
Francillon, J., 4, 5, 76(2), 77
Francis, H., 128(3), 129(3)
Francis, W., 159
Franckcombe, W., 59
Fraser, 173
Fraser, J.M., 120
Fraser, L., 90
Fraser, W.T., 127
Fraser of Lovat, A., 103
Freckleton, Rev.T.W., 140
Frere, Rev.E.H., 131
Frohawk, F.W., 170
Fry, C.E., 131
Furneaux, Rev.A., 136
Fruzer, D., 57
Fysh, Rev.F., 95

G

Gabbitas, Miss E., 169
Gamberini, 179
Gardiner, Mrs. 137
Gardner, G., 92
Gardner, J., 157(4)
Gardner, J.E., 163
Gardner, P.T., 147

INDEX

Gaskoin, 102
Gear, A.T., 164
Geddes, Capt.G., 138
Gerrard, 175
Gibb, 119
Gibb, L., 158
Gibbs, 147
Gibbs, A.E., 155
Giesecke, Sir C.K., 41, 70
Gill, B., 103, 118
Gilles, W.S., 167
Gloyne, C.P., 116(2), 119
Goddard, 143
Godman, F.du C., 158
Goldie, 112
Golding, H., 73
Goldthwaite, O.C., 149
Goodall, 143
Goodall, Rev.J., 86
Goodhall, H.H., 104
Goodman, N., 124
Goodman, O.R. & A.de B., 165
Gore, 122
Gorham, Rev.H.S., 157(2)
Gosling, Rev.W., 60
Goss, H., 147
Gould, J., 114
Gourlie, W., 98
Gowing-Scopes, 169
Grace, T., 59
Graham, 88
Graham, E.W., 172
Graham, R., 21, 22, 88
Graham-Clark, L.T., 133
Grant, Baron, 60
Grant Institute, 174
Grant, J.H., 169(2)
Grant, Gen.S., 155
Grave, R., 70
Graves, P.P., 169
Gray, G.R., 108
Gray, J., 116
Green, 69
Green, J., 164
Green, J.F., 150, 158
Greene, Rev.J., 145
Greenwood, C., 169, 171
Greer, T., 167, 170(2)
Gregson, Col.C.K., 170(2)
Gregson, C.S., 12
Grenville, R., Duke of Buckingham, 41, 43, 90
Greville, Hon.C.F., 179
Griffith, E., 98
Griffith, Rev., 132
Griffiths, G.C., 160
Grist, C.J., 154, 157
Gronovius, 15, 17, 65, 93
Groom-Napier, 130
Grose-Smith, H., 124, 126, 149, 150
Grut, F., 10, 125(3)
Gubba, A.L., 47, 92
Guilding, L., 85, 86
Guthkunst, H.G., 117

Guy, R., 179
Gwatkin-Williams, Capt., 157
Gyngell, W., 161, 163, 164

H

Hadfield, 135(2)
Hadfield, J. & Mrs., 165
Hadley, E.B., 130
Hailes, H.F., 126
Hall, G., 128(2)
Hall, T.W., 157
Hamilton, B., 22
Hamilton, M., 58
Hamilton, W.J., 107
Hammond, L.F., 172
Hammond, W.O., 106
Hammonville, Baron d', 128, 130
Hanbury, F.J., 166
Hanbury, R., 116
Hanson, S., 117, 122
Harford, T., 49, 50, 117
Harker, A., 131
Harper, 108
Harper, J.O., 112
Harper, P.H., 9, 116(2)
Harrington, H., 82(2)
Harris, Rev.G.P., 123
Harris, R.H., 140(2)
Harris, Mrs. S., 116
Harrison, A., 152(2)
Harrison, H.W., 105
Harrison, J., 148
Harting, J.E., 108, 162
Hartley, G.E., 169
Hartwright, J.H., 106
Harwood, B.S., 156, 163, 164, 165
Harwood, W.H., 153(2), 155
Hastings, Marchioness of, 95
Harvey, Rev.R., 106
Hatchett, 102
Haüy, Abbé, 41, 43
Hawke, W., 135
Hawkins, H., 106
Hawkins, Rev.H.H., 134
Hawkins, Rev.H.S., 105
Hawkins, J., 43, 143(2)
Hawkins, T., 35, 88
Hawkins, W., 101
Hawley, Col.B., 169
Haworth, A.H., 4, 6, 83(2)
Haworth, R., 80
Hayes, 179
Haynes, H., 171
Hayward, H.C., 169
Hayward, J.H., 164
Hazeldine, K., 153
Heath, A.E., 148
Hearder, G.J., 128
Hearsey, Brig.Gen.J., 106
Heaviside, J., 81
Helms, R., 126
Helps, J.A., 145
Hemming, A.F., 159

Hemmings, J., 97
Henderson, J., 81
Hennah, Rev.W., 117
Henslow, Rev., 100, 101
Hensman, 127, 128
Heward, R., 21, 105
Hervey, Rev.A.C., 134
Heuland, H., 40, 41, 42, 71, 73, 75(2), 76(2), 78, 79, 80, 81(3), 82(3), 83(3), 84(2), 85(5), 86(2), 87(3), 88(3), 89(2), 90(2), 127
Hewett, W.N.W., 77
Hewitt, Capt.V., 168
Hewitson, 91
Heyde, Rev.T.H., 134
Heyne, A., 139
Heysham, T.C., 98(5)
Heywood, J., 134
Hick, Rev.J.M., 164
Hiden, F.C., 161, 163
Higgins, E.T., 179
Highley, 96
Hill, A., 140
Hill, G.R., 126
Hindson, I., 115
Hislop, R., 139
Hitcham, 179
Hoare, Capt.E., 117
Hodges, A.J., 152
Hodgkinson, J.B., 9, 133(2)
Hodgson, Rev.A.E., 155
Hoey, 112
Holdaway, A.E., 162
Holford, J., 96
Hollis, G., 134
Holmes, Mrs. C., 156
Hood, Lord, 164
Hooker, Sir J., 5
Hope, Mrs. A.E., 175
Hope, J., 167
Hopke, 133
Hopley, E., 106
Horley, W., 148
Horley, W.J., 137
Horley, W.L., 148, 149
Horne, A., 159(3)
Horsfield, T., 99(3)
Horsley, Canon J.W., 158
Horticultural Society, 96, 98
Hoyer, J., 179
Hubbard, J., 100
Hubert, H., 88
Hudd, A.E., 158
Hudson, Mrs., 168
Hudson, W.T., 126
Hughes, 90
Hughes, C.N., 171
Hughes, E., 87
Hughes, R., 59
Huish, 156
Hulkes, 128
Humblot, L., 116
Humphrey, G., 42, 61
Humphrey-Baker, Mrs. D., 171

INDEX

Hunter, 4, 67
Huntley, Marquis of, 115
Hurford, 94
Hurt, C., 94
Hutchinson, J.H., 170

I
Image, Rev.T., 35, 96
Indian & Colonial Exhibition, 119
Ingall, H., 97
Ingall, T., 12, 102(2)
Inman, H.R., 167
Innocent, 70
Inwood, 179
Irby, Col.L.H., 125
Irvine, A., 111
Irving, 143
Irving, Lt.Gen., 103

J
Jackson, 7, 60
Jackson, Mrs., 148
Jackson, J., 65
Jackson, T.W., 161
Jacob, E., 63
Jacobs, Maj.J.J., 164
Jacoby, M., 150
Jager, J., 158
James, Col., 61
James, R.E., 171
Janson, 155
Janvier, 17, 86
Jardine, Sir W., 119
Jeans, J., 58
Jeffrey, W.R., 153
Jeffreys, 136
Jeffreys C., 153
Jenner, 155
Jennings, H.C., 46, 47, 76(2), 77
Jephson, 94
Jerdon, 109
Jerdon, T.C., 16, 92
Jermyn, 180
Jobson, H.W., 148
Johnson, 143
Johnson, C., 114
Johnson, C.F., 163(2)
Johnson, E.E., 164
Johnson, J., 88
Johnson, Miss L.E.M., 173
Johnstone, J., 171
Johnstone, J.F., 169
Joicey, J.J., 150
Jones, A., 172
Jones, A.H., 160(2)
Jones, C.W.B., 162
Jones, E.H., 143
Jones, E.W., 118
Jones, F.W., 155
Jones, T.W., 57
Jourdain, F.C.R., 163
Joy, 156
Joy, E.C., 167
Jupp, B.E., 153

K
Kaltenbach, J.H., 114
Keate, G., 179
Keate, R., 97
Keays, L., 137
Kempt, W., 155
Kempton Manor Hotel, 177
Kennard, M.T., 157
Kent, W.S., 18, 148(2)
Kenward, J., 121
Kilcoursie, Lord, 179
King, 63, 142
King, E.B., 172
King, H.A., 152
King, J., 163
King, T.W., 162
Kirby, Rev.W., 4, 7, 91
Kitchen, V.P., 146
Knaggs, H.G., 107
Knight, J.W., 142
Knight, R., 89
Knighton, Sir W.W., 118
Kock, A., 87(2)
Krapp, 102
Kruper, 103
Kuper, Rev.C.A.F., 119

L
Labrey, B.B., 114, 136
Lacock, 108
Lalande, 62
Lamb, C., 96
Lambert, A.B., 21, 22, 87(3)
Lambert, B., 71
Lambert, C.J., 122
Lang, Rev.H.C., 11, 117, 146
Langdon, A.W., 135
Langstaff, G., 87
Larcom, T.H., 127
Laskey, Capt., 71
Latham, A.G., 112
Latham, J., 74
Lautour, A.de, 99
Layton, T., 154
Lawrence, A., 58
Lawson, 157
Lawson, P., 161
Le Blond, 177
Leach, 176
Leech, J.H., 137, 139
Leake, W., 65
Leathes, 65
Leathes, Rev.G.R., 84(2)
Lee, W., 99
Leeds, A.N., 132
Leeds, H.A., 167
Lees, 135
Leeson, H.B., 109
Le Grand, G.W., 84
Leipner, 131
Leivers, 173
Lester, 164
Lettsom, J.C., 73, 76
Lever, Sir A. (see Leverian Museum)

Leverian Museum, 5, 16, 34, 46, 69(5), 70(3), 72, 80, 82
Lewcock, G.A., 131
Lewis, 121
Lewis, G., 109, 117
Lewis, H., 137
Lightfoot, Rev., 21, 63
Lilford, 147, 148
Limminghe, Comte de, 101
Lindemann, O., 161
Linden, A., 119(2)
Lindley, 103
Lingwood, H., 145
Linnaeus, 159
Linnean Gallery, 74
Linnean Society, 102
Linnell, J., 149
Linter, Miss, 149
Liptrap, J., 68
Lisle, Commodore, 45, 57
Little, B., 66
Livett, H.W., 138
Livingstone, C., 119, 124
Lloyd (see Llwydius)
Lloyd, A., 160
Lloyd, L.R.W., 166
Llwydius, 75
Loat, W.L.S., 153
Lobb, J., 172
Locke, 135
Longhurst, A.M., 163
Longley, H., 90
Longley, W., 129
Lord, J.K., 102
Lovell-Keays, A., 136
Low, 113
Low, J.S., 118
Low, W.H., 136
Lowther, R.C., 170(2)
Ludlam, G.S., 162
Lwyd, 82
Lynch, C., 100
Lynch Piozzi, Mrs. H., 79
Lynn, 62

M
Mabille, J., 154
Macdonald, K.C., 141
Machin, W., 9, 129, 130
MacLeay, A., x, 4, 6, 74
Maddison, J., 11, 72
Maddison, T., 148(2)
Main, 168, 170
Main, H., 152
Mainwaring, Gen.G.B., 50, 129
Maitland, F.L., 153
Malcolm, Col.G., 172
Maling, W., 151
Manders, Col.N., 155
Manger, 155
Mannering, Rev.E., 168
Manners, Lord, 64
Mann, 105
Mantell, G.A., 93

INDEX

Mantua, Duchess of, 130(6), 131, 132(3)
Marchant, G.le, 164
Marcon, Rev.J.N., 13, 167, 169(5)
Marlow, J., 57
Marsden, H.W., 129, 136, 152(2)
Marsh, H., 162
Marshall, E., 163
Marshall, E., 179
Marshall, Rev.F., 138
Marshall, J.T., 154
Marshall, W., 109, 123, 124, 157
Marsham, R., 111
Marsham, T., 5, 6, 78
Martin, K.B., 98
Martyn, 69
Mason, P.B., 11, 141, 142(2), 143(2), 144, 147
Massey, H., 137, 166(2)
Mathew, G.F., 121(2), 145, 153, 154(2)
Matthews, Rev.A., 134, 137
Matthews, Rev. T., 62
Matthias, H.W., 134
May, H., 160
May, J.W., 140
Mayne, D., 58
McArthur, H., 149
McCardle, L.F., 175
McClelland, H., 166
McDonald, R.J., 176
McEnery, Rev.J., 36, 87
McLachlan, R., 166
McLeod, Sir M., 168, 170(3)
Mead, R., 57
Meaden, L., 163
Meek, 119
Meek, E.G., 113, 118, 125
Melvill, J.C., 48, 132, 135, 142, 163
Menge, 180
Menish, H., 72
Mercer, A.H., 108
Merrin, J., 136
Metcalf, 102
Michaud, J., 173
Middlebrook, 147
Midgley, 180
Milburn, Maj.W.H., 158
Millais, J.G., 147
Millan, J., 61
Miller, T., 91
Milligan, J., 117
Milne, G., 4, 6, 79, 86
Milner, Sir W., 129
Mills, H.O., 154
Mills, Y.H., 148
Milton, 92
Minard, Mrs., 64
Mitford, R.H., 120
Mitford, R.S., 156
Mitchell, A.T., 165
Mitchell, F.S., 124
Mivart, St.G., 109

Mohr, 66
Möller, 126
Möller, O., 137
Moore, 89, 111, 128(2)
Moore, F., 146
Moore, Maj.F.C., 161
Morley, W.A., 119, 156
Morris, C.H., 145
Morris, Rev.F.O., 10, 127, 149
Morris, J.B., 165
Morris, R., 61
Morton, C., 66
Mosse, G.S., 119
Motherwell, 84
Mounsey, J.D., 172
Munn, H., 96
Munn, P.W., 154
Munt, H., 136, 139
Murray, 92
Murray, A., 8, 112, 151
Murray, Rev.R.P., 118
Murray, W.C., 113
Mutch, J.P., 164
Mylne, G. (see Milne)

N

Naish, A., 125
Nash, G., 165
Nash, J., 108
Neck, J.F., 150
Neilson, J., 62(2)
Nelson's Pharmacy, 176
Nevill, H., 50, 140
Neville, Lady D., 112
Nevinson, B.G., 151
Nevinson, E.B., 161, 163
Newan, W., 137
Newcombe, S.P., 152
Newey, H.F., 171
Newman, L.H., 171, 172
Newnham, C.E., 164
Newnham, F.B., 159
Nichols, J.B., 162
Nickels, W.L., 157
Nicoll, W., 106
Nicolson, W.E., 153
Noakes, A., 155
Noble, H., 130, 134, 135(2), 137, 139, 141, 142(2), 143, 157, 165
Norgate, F., 156
Norris, T., 49, 109(4)
Northcote, T., 58
Northumberland, Duke of, 43, 174(2)
Norton, Mrs. R., 170
Norwegian Expedition, 141
Nowers, J.E., 127
Nurse, Lt.Col.C.G., 164, 169
Nuttall, 99

O

Odd, D.A., 175
Ogilvie-Grant, W.R., 157
Olwer, G.B., 149, 156, 172

O'Reilly, T., 129
Ormerod, Miss E.A., 138, 155
Osborne, A.P., 134
Owen, A., 60
Owen, T.B.B., 105
Owston, A., 153

P

Packman, 120
Paddock, G.F., 155
Page, W., 128
Page, H.E., 169
Palliser, H.G., 149
Pamplin, W., 102
Pankhurst, 156
Pardoe, J., 148
Park, M., 17
Parke, S., 180
Parker, Sir W., 176
Parkinson, J., 35, 36, 80, 105
Parlby, Rev.S., 68
Parlett, J., 177
Parris, R.S., 163
Parritt, H.W., 163
Parry, 118, 119, 130
Parry, F.J.S., 118(2)
Parry, T., 109
Parsons, J., 64
Partridge, C.E., 149
Patterson, G., 134
Paul, Mrs. 107
Payne, J.H., 83
Peach, A.W., 163
Pearman, Capt.A., 165
Pearson, 66
Pearson, H., 126
Peek, Sir W., 162
Penn, W., 169
Pennington, F., 167
Penrose, Rev.T., 61
Percy, J., 123(4)
Pereira, 96
Perkins, A., 108
Perkins, V.R., 158
Perkins, W., 105
Pether, W.G., 165
Philips, J.L., 75
Phillips, Capt., 167
Phillips, Rev.E.J.M., 101
Phillips, R., 92
Pickard, L., 67
Pickett, C.P., 167
Pickles, F., 163
Pidsley, W.E.H., 144, 145(3), 158
Pierson, Archdeacon, 70
Pinder, 115, 142
Piozzi (see Lynch Piozzi)
Pitchard, G.B., 139
Pitman, 156
Playfair, J., 78
Plympton, R., 79
Pococke, 58
Pond, A., 58, 62
Popham, F.L., 114

INDEX

Porritt, G.T., 162
Portland, Duchess of, 4, 21, 46, 62
Potts, T.H., 94(2)
Pratt, 128
Pratt, C., 152
Pratt, F., 152
Prest, E.E.B., 159
Prest, W., 103, 115, 117
Preston, Rev.D., 126, 149
Preston, E.S., 98
Preston, H.B., 154
Price, L., 164
Proctor, Maj.F.W., 152, 153, 155
Proctor, W., 97
Prosser, R., 95
Proudfoot, Rev.S., 164
Prout, L.B., 148
Pruen, J.A., 130, 132, 134
Pulteney, 102
Purnell, B.P., 108

Q

Quekett, J.T., 101
Quibell, W., 164, 173

R

Raben-Levetzau, Baron, 25, 175
Rackstrow, 66(2)
Raddon, W., 81
Radley, P., 162
Raine, F., 117, 129
Ramsden, H., 145
Randall, 81
Rattray, Col.R.H., 150(2), 151, 153(2)
Rawle, W., 64
Rawson, Sir H.R., 136, 140
Raynor, Rev.G.H., 12, 125, 146(2), 150, 151, 153, 154, 155, 160, 161, 163(2)
Rayward, A.L., 165
Read, Rev.G.R., 103
Reed, A.D., 169
Reed, J.T.T., 151
Reeve, L., 102, 104(2)
Reeves, W.W., 126
Rehe, S., 70
Reid, H., 103
Reid, P.C., 162
Reid, Capt.S.G., 125
Reis, A., 124
Remy, 140
Rennie, W., 64
Renton, W., 148
Reuss, C.K., 165
Reyne, 110
Reynolds, F.A., 110
Richards, Rev.F.M., 171
Richards, P., 155
Richardson, J., 175
Richmond, Duke of, 57
Riding, W.S., 151
Ridley, P.W., 162, 164

Rimmington, J.W., 124(2)
Rivers, Lord, 81
Robertson, C.M., 156
Robillard, V.de, 49
Robson, Col., 70
Robson, J.E., 10, 130(2), 146
Robson, J.P., 172
Rogers, E.A., 152
Rooke, W.D., 120
Roper, 63
Roper, J., 61
Rosa, L.G., 13, 174
Rose, 51
Rosehill, 167
Rosenberg, 154
Rosevear, J.B., 158
Rosomon, H.jun., 136
Ross, Gen., 91
Ross, H., 131
Rosser, C.W.S., 128
Rouby, J.J., 59
Rouse, G., 126
Rowberry, D., 174
Rowe, R., 91
Rowlands, J.F., 163
Rowley, G.D., 164
Royal Museum, 78
Royal United Services Institute, 100(2)
Rucker, S., 49, 109, 113
Ruddy, Mrs., 171
Rumbold, Sir T., 65
Ruskin, 175
Ruskin, J., 43, 116
Russ, P., 131
Russell, G.M., 172
Russell, J.W., 158
Russell, S.G.C., 171, 172(2)
Russell, W., 117
Russell, W.T., 103
Rustafjaell, R.de, 147, 153
Ryder, G.R., 138
Rylands, P.H., 122

S

Sadler, T., 57
St.Bartholomew's Hospital, 161
St.John, Rev.S., 127
St.John, W., 162
Sale, R.C., 88
Sall, J., 121
Salmon, J.D., 100
Salperton Park, 118
Salwey, R.E., 125
Sang, J., 8, 115(2), 120
Sargent, F., 100, 104(2)
Saull, 102
Saunders, E., 150
Saunders, G.S., 150
Saunders, H., 111, 125
Saunders, Sir S.S., 117
Saunders, W.W., 10, 110, 111
Sauze, A., 155
Saxby, 110

Schill, 135
Schill, C.H., 148(4), 151
Schleicker, M., 92
Schmassmann, 154(2), 155
Schomburgk, Sir R., 88, 103
Schooling, H.C., 123
Scollick, A.J., 155
Scorer, A.G., 163
Scott, J., 107, 108
Scott, J.W., 160
Scott, M.N.D., 169
Seale, R.F., 86
Sealy, A.F., 104
Sequeira, J.S., 145, 153
Seymer, H., 180
Shannon, D.S.B., 168
Sharman, F.W., 171
Shearwood, G.P., 117, 125
Sheffield, 101
Sheldon, T., 68
Sheldon, W.G., 168(2)
Shelley, Capt.G.E., 122
Shepherd, A.H., 147
Shepherd, E., 127
Shepherd, E.R., 116
Shepherd, J., 171
Sheppard, A.F., 121
Sheppard, Maj.E., 121
Sheppard, E., 140
Sherwill, J.L., 121(2)
Shirley, 135(2)
Shirley, S.E., 138, 139, 143
Shuckard, W.E., 8, 102
Silver, Rev.F., 117
Simmons, C.H., 153
Simpson, T., 150, 151
Sims, J., 21, 81
Sinfield, 89
Sirr, Maj.H., 81, 90
Sivright, T., 84
Skinner, E.R., 150
Skinner, K.L., 149
Skinner, P.F., 162
Sladen, C.A., 131
Sladen, E.H.M., 131
Slater, Rev.H.A., 153
Slater, J., 84
Slater, P.H., 170
Sleath, 67
Slocombe, S., 162, 163
Smalley, F.W., 160
Smallpiece, A.M., 148
Smart, Rev.G., 133
Smart, H.D., 168, 169
Smart, J., 132
Smee, Gen., 112
Smith, C., 123
Smith, F., 8, 113
Smith, H.B., 169
Smith, J., 132
Smith, M., 73
Smith, S., 105, 126
Smith, S.G., 172(3), 173
Smith, Rev.U., 121(2)
Smith, W.H., 154(2)

INDEX

Smith, W.P., 94
Smithe, Rev.F., 143
Smyth, Lady G., 152
Snowden, Canon A.H., 164
Solander, D.C., 4, 180
South, R., 134, 135
Southgate, Rev.R., 65
Sowels, F., 173
Sowerby, G.B., 47, 48, 49, 82(2), 83, 97
Speed, 62
Sperring, C.W., 164
Spiller, A.J., 127
Stafford, W., 123
Stainton, H.T., 135
Standen, R.S., 126, 153
Standish, 131
Standish, F.O., 113
Standish, J., 108
Stanley, E.S., 92
Stark, A.L., 139
Steele, E., 57
Stephens, A., 152
Stephens, J.F., 6, 79, 135
Stephenson, E., 117
Stephenson, R., 113
Stevens, J., 71
Stevens, O.C., 176
Stevens, S., 10, 136(2)
Stevenson, H., 162
Stewart, 147(2)
Stiff, Rev.A., 166
Stiff, P., 167, 168, 169
Stiff, P.A., 172
Stoate, W., 146, 147
Stokes, C., 94
Stoke-Roberts, Mrs., 128
Storer, G.H., 174
Stothard, T., 93
Stowell, Rev.H., 119
Strange, J., 180
Streatfield, J.F., 98
Strickland, C., 177
Strong, C.E., 124
Stuart, Mrs., 115
Stuart, Gen.C., 81
Stuart, J. (see Bute)
Stuart, Mrs. W., 133
Stubbs, Rev.C., 154
Stukeley, W., 58
Sturges, J., 59
Sturt, F., 147
Stutchbury, H.R., 88, 91
Stutchbury, S., 80, 99, 100
Styan, T.G., 160
Sullivan, Sir R., 71
Sumner, J., 68
Surrey Museum, 84
Surrey Zoo, 95
Sutton, G.P., 172
Swaffham Prior Natural History Society, 180
Swainson, W., 4, 6, 7, 47, 79, 86
Swansea, Lord, 156
Swinburne, H., 68

Swinhoe, E., 126, 127, 128, 135, 136, 158
Syme, J.B., 123

T

Tait, R., 168
Talbot, Sir C., 74
Tamnau, F., 88
Tancock, Canon O.W., 152, 163
Tankerville, Earl of, 84, 180
Tassie, J., 66
Tautz, P.H., 155
Taylor, Mrs., 127
Taylor, H., 94
Taylor, Col.J.E.C., 170
Taylor, T.L., 48, 49, 113, 162
Taylor, W.G., 152
Teesdale, 92
Telford, A.E., 176
Tempany, Mrs., 152
Tennant, 153
Tennant, J., 43, 176
Terry, Maj.H.A., 159
Terry, W., 117(2)
Thackeray, Provost, 135
Thatcher, Mrs., 148
Theobald, J., 59
Thompson, J.A., 160, 170(2)
Thornewill, Rev.C.F., 160
Thornhill, E.H., 159
Thornthwaite, W.H.E., 148(2)
Thornton (see Linnean Gallery)
Threlfall, J.H., 154
Thurnall, C., 129
Todd, R.G., 158, 172
Tomkins, J., 111
Touch, J.de la, 133
Townsend, Rev., 77
Trattle, M., 82
Travers, 139
Tristram, H.B., 94, 97(2), 99, 103
Troughton, N., 106
Truman, J., 94
Tucker, 91
Tucker, M., 132
Tugwell, W.H., 131(2)
Tuke, 80
Tuke, J.H., 131
Tunaley, H., 167
Tunstall, W., 147
Turle, W.H., 147
Turner, E., 150, 157
Turner, H.J., 168
Turner, J.A., 101, 114
Turner-Bridger, R.C., 174
Tutt, J.W., 150, 151(2), 152(2)
Tyler, Capt.C., 131, 132(2)

U

Urwick, W.F., 143, 145, 146
Ushaw College, 174
Ussel, Madame, 148
Ussher, R.J., 139

V

Valentine, A., 163, 173
Vaughan, H., 9, 123(2), 126
Verel, J.B., 162
Verkreutzen, 119
Vernède, H., 47, 99
Vickers, J., 162
Villet, M., 80
Vingoe, 121, 122
Vipan, A.M., 168
Virtue-Tebbs, H., 136
Von Ruprecht, 70

W

Wailes, G., 10, 117
Walker, 64, 66
Walker, E., 100
Walker, E.J., 119
Walker, F., 6, 111
Walker, Sir P., 6, 85
Walker, S., 167
Wallace, 120
Wallace, A.R., 108, 121
Waller, A., 123(2)
Wallis, G., 111
Wallis, R., 139
Wallis, S.H., 137, 138, 142, 144, 146
Wall-Row, J., 164
Wall-Row, T., 155
Walter, H.F., 152
Walton, J., 101
Warburton, 7, 88
Ward, D., 175
Ward, D.F., 169
Ward, E.H., 111
Ward, O., 116
Waring, 113
Waring, S.L., 112
Warne, W.F., 154
Warren, W., 120
Warszeweiz, A., 91
Warwick Castle, 174
Waterhouse, E.A., 154
Waterhouse, F.H., 154
Waterhouse, G.R., 113
Watkins, W., 121
Watkins & Tullett, 158
Watkinson, Canon G., 173(2)
Watson, A., 171
Watson, D., 170
Watson, J., 102, 107
Watson, Rev.J., 128
Watts-Russell, J., 110
Weatherhead, 89
Weaver, T., 95
Webb, S., 12, 135, 156(2), 157(3)
Webber, W., 72
Webster, 151
Webster, T., 35, 89
Weir, J.J., 128(3)
Welch, Mrs. A., 72
Welcome, 167
Wellman, J.R., 128

INDEX

Westland, W., 154
Westminster Aquarium, 112
Weston, Rev.S., 81
Weston, W.P., 114
Wharton, H.T., 141(2)
Wheeler, E., 153, 154
Wheeler, F.D., 130(2)
Wheelwright, H.W., 100, 101(2), 103(2), 104
Whitaker, J., 123
White, E.B., 171
White, H.J., 140
White, J., 63, 73
White, S., 118
Whitehill, Col., 89
Whitehouse, Sir B., 13, 168(5)
Whiteley, 135
Whitfield, R.G., 111
Whymper, E., 152
Whyte, 112
Wickham, Rev., 165
Wilde, J.P., 145
Wilkin, S., 5, 76
Wilkinson, 112
Wilkinson, Rev.C., 139, 140
Wilkinson, S.J., 138
Wilkinson, W., 171

Wilkinson, W.A., 154, 164
Willett, R., 74(2)
Williams, 98, 132
Williams, C.H., 165
Williams, H.B., 171(3)
Williams, J.M., 161
Williamson, J.B., 140
Willmot, C., 149
Willshee, C.J., 172
Wilson, J., 146
Wilson, W., 72, 110
Windeler, D., 85
Wing, W., 96
Winstone, G.D., 164, 172
Wise, Mrs., 120
Wise, J.R., 93
Wollaston, J.V., 128
Wollaston, T.V., 136, 151
Wolley, J., 93, 94, 95, 96, 97(2), 98, 100, 101, 102, 152
Wombwell, 108
Wood, C.R., 163
Wood, H., 168
Wood, J., 123
Woodcock, 120
Woodd, J., 74
Woodford, Archdeacon, 70

Woodford, C.M., 122(2), 149
Woodford, E.J.A., 72
Woodforde, F.C., 141
Woodroffe, Rev.D., 155
Woods, W.G., 109
Woodward, J., 17, 34, 57, 82
Woollett, G.F.C., 173
Woscott, Rev.J., 66
Wright, 78
Wright, A.E., 170(2)
Wright, B.M., 110, 120, 144
Wright, Rev.H.J., 163
Wright, W.C., 145
Wynn, G.W., 143

Y
Yarrell, W, 96
Yeats, T.P., 4, 61, 62
York, Col.P.J., 110(2)
Yorkshire Museum, 117
Young, J., 138(2)
Young, M., 87
Young, S., 131

Z
Zoffany, J., 73
Zouche, Lady, 120